AKADEMIE DER WISSENSCHAFTEN DER DDR

Zentralinstitut für Physik der Erde (ZIPE)

Seismological Bulletin 1973 Station Moxa (MOX)

By

Johannes Stelzner, Dorothea Güth, and Joachim Weyrauch

With 1 Figure

AKADEMIE-VERLAG · BERLIN
1978

Erschienen im Akademie-Verlag, 108 Berlin, Leipziger Straße 3–4

Lizenznummer: 202 · 100/470/78
Gesamtherstellung: Druckerei „Thomas Müntzer“, DDR-582 Bad Langensalza
Bestellnummer: 7626780 (2004/B/1973) · LSV 1437
Printed in GDR
DDR 39,— M

PREFACE

The annual Seismological Bulletin 1973 for the Station Moxa (MOX) continues the series of publications about seismological records at the Central Station Moxa edited by the Central Earth Physics Institute of the Academy of Sciences of the German Democratic Republic.

The provisional analysis of the records of station Moxa was performed in the Seismological Service under the direction of JOHANNES STELZNER by JOACHIM WEYRAUCH and BRIGITTE HÄNSCH.

The annual Bulletin 1973 was prepared by JOHANNES STELZNER, JOACHIM WEYRAUCH and DOROTHEA GÜTH with the technical assistance of URSULA DÖRING.

Control of the instruments of the station Moxa was carried out under the direction of CHRISTIAN TEUPSER.

H. KAUTZLEBEN
Director

TABLE OF CONTENTS

Preface 3

Table of Contents 5

The Seismological Bulletin

Preliminary Notes for Interpretation of Seismograms 7
Seismographs of Station Moxa and their Parameters 1973 . . . 11
Amplitude Characteristics of Station Moxa 1973 13
Seismological Recordings at Station Moxa 1973 15

Preliminary notes for the interpretation of seismograms

In the Bulletin the international code is used:

1. Phase interpretation

Pg — direct longitudinal wave in near epicentral distances ($D < 10°$)

Pb, Pn — guided longitudinal head waves along the CONRAD- or MOHOROVIČIĆ-discontinuity ($D < 10°$)

P — direct longitudinal wave travelled through the earth mantle

P diff — direct longitudinal wave diffracted around the core boundary

PKIKP — direct longitudinal wave travelled through the inner core (traved-time branch DF)

PKHKP — direct longitudinal wave refracted in the intermediary zone between inner and outer core. Phase symbol according to BOLT [1] (traveltime branch GH)

PKP2 — direct longitudinal wave travelled through the outer core only (travel-time branch AB)

PKP — first noticeable onset of longitudinal core phase not identified

PP, PPP — waves reflected at the earth surface with permanent longitudinal character

PKKP — core phase reflected once within the core at the outer core boundary

PKPPKP — longitudinal core phase reflected at the earth surface

Sg — direct transversal wave in near epicentral distances ($D < 10°$)

Sb, Sn — guided transversal head waves along the CONRAD- or MOHOROVIČIĆ-discontinuity ($D < 10°$)

S — direct transversal wave travelled through the earth mantle

SKS — direct wave travelled transversal through the mantle and longitudinal through the core

SS, SSS — waves reflected at the earth surface with permanent transversal character

SKKS — wave travelled transversal through the mantle, longitudinal through the core and reflected within the core at the outer core boundary

PcP, ScS, PcS, ScP — longitudinal and transversal waves with steady or changing character reflected at the outer core

PS, SP, PPS — longitudinal and transversal waves with changing character reflected at the surface of the earth

pP, sP. pPP, sPP, pPKIKP, sPKP2, pS — phases of deep-focus earthquakes of longitudinal or transversal waves with steady of changing character. p;s — reflected near the epicentre

pPKP, sPKP — phases of deep focus earthquakes of longitudinal core waves not exactly to be coordinated

SKP, PKS — core phases with different character before and after the direct transit of the core

SKSP — SKS wave with longitudinal character after the reflection at the surface of the earth

P1, P2, P3,. . ., S1, S2, . . . — multiple onsets of body waves

Pn, Sn — teleseismic Pn and Sn waves in the epicentral distances $23° < D < 40°$ after BÅTH [2]

Pa, Sa — waves probably guided in the astenosphere channel or higher modes of surface waves

PL — leaking modes, normal dispersed train of waves of periods greater than about 10 s, beginning at or near the time of initial P-wave

X, Y, Z — remarkable phases of body waves, not to be identified

LmV, LmH — maximum of the vertical and horizontal component respectively of longperiodical surface waves. If there are several maxima with comparable proportions in A/T, the numeration was carried out in a temporal sequence e.g. Lm1H, Lm2H

The phase symbol is followed by the designation of the type of seismometer from which the time of onsets is taken.

A — seismograph with amplitude characteristic of type A (short-period)

B — seismograph with amplitude characteristic of type B (middle-period)

C — seismograph with amplitude characteristic of type C (long-period)

2. Measurements of amplitudes and calculation of magnitudes

All data of amplitudes and periods printed in the column "remarks" are always taken from the records of the same instruments, from which are taken the onset-times of the corresponding phases. The symbol of phase and component is followed by the symbol of the type of instruments e.g.: PV A, PV B, LmH B, LmV C.

Data of amplitudes obtained from records of instruments of type A are given in units of length of nm (1 nm = 1 nanometre = 10^{-6} millimetre). Data of amplitudes obtained from instruments of type B and such obtained from instruments of type C are given in units of length μm (1 μm = 1 micrometre = 10^{-3} millimetre) e. g.: PVA 1.3 s 38.6 nm, SHB 10 s 3.2 μm, LmH B 22 s 15 μm.

Magnitudes are determined from all those phases, for which calibrating functions are known and internationally used, i. e.

for maxima of body waves P(PH, PV), PP(PPH, PPV), and S(SH)-Q-functions from Gutenberg and Richter [3] — and

for maxima of surface waves ($h < 100$ km) LmH, LmV — calibrating functions from Prague σ [4] —.

The station correction S was not yet taken into consideration.

MB — magnitude of vertical component V of the first onset of P-waves given by Neic

MS — magnitude of horizontal component H of the maximum surface wave given by Neic

M — magnitude calculated from given data of station Moxa. Notice the wave type and the type of instruments written on the same line

3. Direction of body-wave onsets

If the direction of motion at the beginning of a wave onset is clearly to be recognized, the sign + or — is placed before the phase symbol. It means:

in the Z component	+	ground motion upwards, compression
	—	ground motion downwards, dilatation
in the N component	+	ground motion to the north
	—	ground motion to the south
in the E component	+	ground motion to the east
	—	ground motion to the west

4. Further abbreviations

i — sharp beginning of phase motion (impetus)

e — gradual beginning of phase motion (emersio)

D — epicentral distances in degree (°), calculated according to geocentric coordinates, the maximum error of the own calculations amounts to ± 0,1°.

Az — azimuth: clockwise measured angle between north direction in epicentre and the connecting line from epicentre to station Moxa

h — depth of focus in km, our data for depth of focus are based on travel-time curves for deep focus earthquakes after GUTENBERG and RICHTER [5]

H — orgin time in UTC (Universal Time)

NEIC — National Earthquake Information Center, Boulder, Colorado, USA (until May 1973)

NEIS — National Earthquake Information Service, Boulder, Colorado, USA (from June 1973)

BCIS — Bureau Central International de Sêismologie Strasbourg, France,

ANUSSR — Akademia Nauk USSR, Moscow

AEC — United States Atomic Energy Commission, Washington, D. C.

ISC — International Seismological Centre, Edinburgh

Hannover — Niedersächsisches Landesamt für Bodenforschung Hannover-Buchholz, FRG

For abbreviations of seismological stations and other agencies in the international three letter code see the introductions to the Regional Catalogue of Earthquakes, Edinburgh and the Bulletins of the International Seismological Centre, Edinburgh.

Round brackets indicate uncertainties in interpretation of phase, time, depth of focus or epicentral distances, respectively.

[1] BOLT, A., The velocity of seismic waves near the earths center. Bull. Seism. Soc. Am. 54 (1964) 1, 191—208.

[2] BÅTH, M., Propagation of Sn and Pn teleseismic distances. Pure and Applied Geophysics 65 (1966/II) 19—30.

[3] GUTENBERG, B. and RICHTER, C. F., Magnitude and energy of earthquakes. Annali di Geofisica 9 (1956) 1, 1—15.

[4] KÁRNÍK, V., KONDORSKAJA, N. V. u. a., Standardization of the earthquake magnitude scale. Stud. Geophys. et Geodet., Prague 6 (1962) 41—48.

[5] GUTENBERG, B. and RICHTER, C. F., Materials for the study of deep-focus earthquakes. Bull. Seism. Soc. Am. 26 (1936) 4, 341—390.

Seismological Station Moxa (MOX) of the Central Earth Physics Institute

Elevation above mean sea level:	455 m
Bedrock:	clay slate of the lower carboniferous formation
Geographic coordinates:	$\varphi = 50°38'46''N$ $\lambda = 11°36'58''E$
Address:	Central Earth Physics Institute Seismological Service DDR-69, Jena, Burgweg 11 German Democratic Republic Telex: 05886275 seis dd

Seismographs and their parameters 1973

T_s — seismometer free period
T_g — galvanometer free period
D_s — seismograph damping
D_g — galvanometer damping
V_0 — magnification factor
N — north-south component
E — east-west component
Z — vertical component
σ^2 — coupling coefficient

SKM — Seismograph Kirnos modified
SSJ — Seismic Station Apparatus Type Jena
VSJ — Vertical Seismograph Type Jena

Type of Seismograph	Comp.	T_s/[s]	T_g/[s]	D_s	D_g	V_0	σ^2
VSJ II	Z	0.23	0.065	0.33	1.2	300000	0.048
VSJ II	Z	1.0	1.0	0.5	0.5	47200	0.56
A SKM III	N	1.64	0.39	0.52	1.98	25700	0.047
	E	1.64	0.39	0.51	1.93	25400	0.046
until May, 17.	Z	1.64	0.39	0.51	1.99	24600	0.050
A SKM III	N	1.64	0.39	0.52	1.97	24000	0.047
from May, 19.	E	1.63	0.40	0.50	1.92	24700	0.047
	Z	1.64	0.39	0.51	1.96	23400	0.050
B SSJ I	N	20	1.13	0.50	8.87	109	0.077
		20	1.14	0.50	8.79	1110	0.077
	E	20	1.13	0.49	8.85	103	0.070
		20	1.16	0.49	8.61	1050	0.070
	Z	20	1.13	0.48	8.82	108	0.057
		20	1.24	0.48	8.05	910	0.057
C SSJ I/L	N	30	89.5	1.49	0.5	1000	0.085
	E	30	74.4	1.24	0.5	1010	0.051
	Z	30	91.0	1.52	0.5	1060	0.108
Strain/L	N		85.4		0.70	65*	
	E		86.2		0.70	67*	
(coupled)	N + E		86.2		0.70	42*	

*) for apparent wave velocity 5 km · s^{-1}

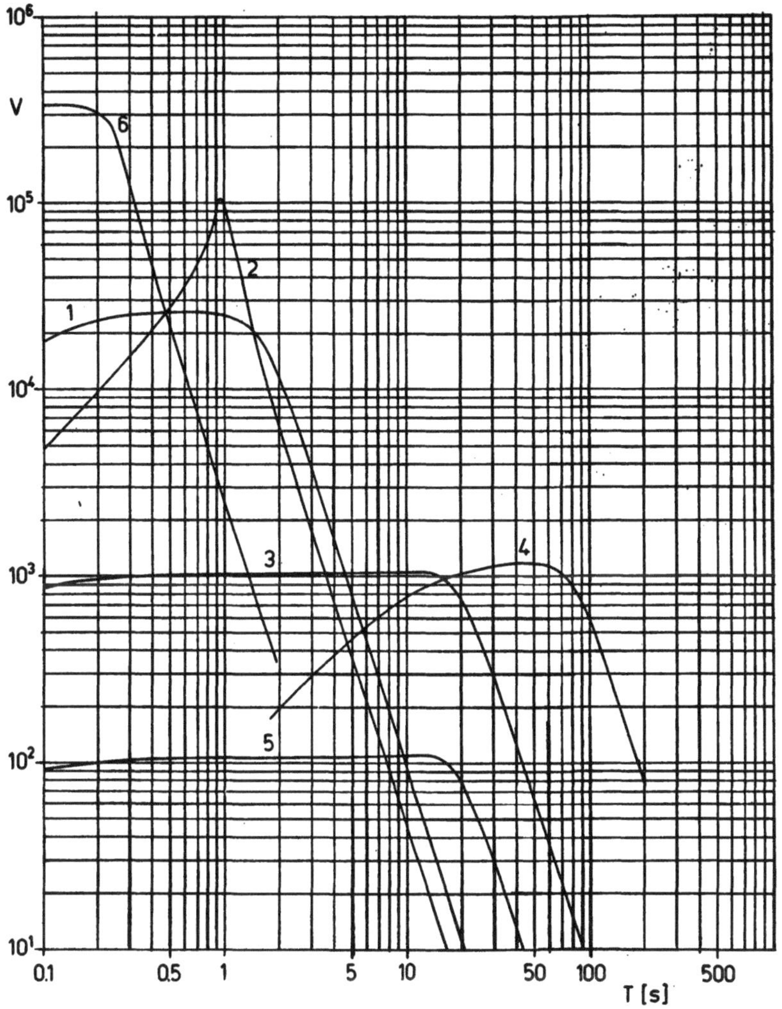

Fig. 1. Station Moxa, mean amplitude characteristics 1973

1 — Seismograph Kirnos Modified-III (SKM-III) (NS-, EW- and Z-component)

2 — Seismograph Type Jena II (Z-component)

3 — Seismic Station Apparatus Type Jena I/1000 (SSJ-I/1000) (NS-, EW- and Z-component)

4 — Seismic Station Apparatus Type Jena I/L (SSJ-I/L) (NS-, EW- and Z-component)

5 — Seismic Station Apparatus Type Jena I/100) (SSJ-I/100) (NS-, EW- and Z-component)

6 — Seismograph Type Jena II (Z-component)

Seismological Recordings at Station Moxa 1973

Day	Phase		h m s	Remarks
1.	ePKP	A	04 05 10	East New Guinea Region
	e	A	05 14.5	9.21 S 150.63 E
				H = 03 46 09.8 h = 41 km MB = 5.3
				D = 126.67 Az = 328.6 (NEIC)
1.	ePKP	A	05 42 02	Tonga Islands 15.01 S 193.96 W
				H = 05 22 29.8 h = normal MB = 5.0
				D = 144.17 Az = 353.9 (NEIC)
				PKPV A 1.7s 42.4nm
1.	eP	A	11 55 33	South Atlantic Ridge 35.51 S 16.21 W
	ePP	A	58 54.5	H = 11 42 37.5 h = normal MB=5.4 MS=6.0
	eS	C	12 06 12	D = 89.23 Az = 17.3 (NEIC)
	ePS	C	07 25	LmH B 16s 2.9/um M = 5.8
	eSS	BC	12 00	LmV B 17 3.5/um 5.9
	LmH	B	34.0	
	LmV	B	34.1	
2.	eP	A	01 53 44.5	Greece 38.22 N 20.17 E
	e	A	53 56	H = 01 50 28.5 h = normal MB = 4.3
				D = 13.83 Az = 336.7 (NEIC)
2.	LmH	C	01 55.0	Sumbawa Islands Region 9.85 S 117.43 E
				H = 00 53 20.3 h = 66 km MB = 5.5 (NEIC)
				D = 107.58
2.	e	A	02 41 15	Molucca Passage 1.04 N 126.21 E
				H = 02 27 09.2 h = 61 km MB = 5.4
				D = 104.54 Az = 323.3 (NEIC)
				traces
2.	eP	A	04 00 46.5	South of Panama 5.40 N 82.54 W
	LmV	B	33.0	H = 03 47 52.5 h = 30 km MB = 5.2
	LmH	B	35.0	D = 88.50 Az = 39.4 (NEIC)
2.	eP	A	22 35 51.5	Tibet 31.24 N 88.09 E
				H = 22 25 57.0 h = normal MB = 5.2
				D = 58.30 Az = 313.3 (NEIC)
				PV A 1.8s 43.9nm M = 5.3

Day	Phase		h m s	Remarks
2.	eP	A	23 25 14.5	Jan Mayen Island Region
				71.33 N 7.55 W
				H = 23 20 16.7 h = normal MB = 4.7
				D = 22.52 Az = 146.9 (NEIC)
				PV A 1.3s 32.8nm M = 4.6
3.	eP	A	03 11 13	Santiago del Estero Prov., Argentina
	ePP	A	15 30	27.72 S 63.26 W
				H = 02 58 16.7 h = 563 km MB = 5.6
				D = 102.09 Az = 38.9 (NEIC)
3.	+eiP	A	14 39 03.5	Tadzhik SSR 39.11 N 71.89 E
	ePP	A	40 48.5	H = 14 31 04.5 h = normal MB = 5.5
	eSS	B	48 55	D = 43.10 Az = 306.0 (NEIC)
	LmV	B	15 00.9	PV A 1.5s 70.3nm M = 5.2
	LmH	B	01.0	LmH B 12 1.0 μm 4.9
				LmV B 14 1.6 μm 5.2
3.	eP	A	21 26 23.5	Off Coast of Oregon 44.34 N 129.15 W
	LmH	C	22 04.0	H = 21 14 16.4 h = 18 km MB = 5.3
	LmV	C	04.0	D = 79.44 Az = 24.2 (NEIC)
				PV A 2.0s 59.8nm M = 5.3
4.	ePKIKP	A	01 26 44.5	New Hebrides Islands 13.38 S 167.09 E
	e	A	26 53	H = 01 07 50.3 h = 194 km MB = 5.1
	ePP	A	29 43	D = 137.82 Az = 336.8 (NEIC)
4.	eP	A	03 35 36	Jan Mayen Island Region 71.59 N 6.70 W
				H = 03 30 38.1 h = 33 km MB = 4.4
				D = 22.59 Az = 148.6 (NEIC)
				PV A 1.4s 23.2nm M = 4.5
4.	eP1	AB	08 08 46.5	Jan Mayen Island Region 71.11 N 7.67 W
	eP2	A	08 51	H = 08 03 50.4 h = normal MB = 5.1
	eS	B	13 00	D = 22.36 Az = 146.4 (NEIC)
	LmH	B	19.4	P1V A 1.2s 48.7nm M = 4.8
	LmV	B	20.0	P2V A 1.2 114.0nm 5.2
				LmH B 13.5 1.9 μm 4.7
				LmV B 13 1.6 μm 4.8

Day	Phase		h m s	Remarks
4.	eP	A	12 08 13.5	Jan Mayen Island Region 71.57 N 7.28 W
				H = 12 03 13.9 h = normal MB = 4.2
				D = 22.68 Az = 147.7 (NEIC)
5.	+eiP	A	01 49 52	North Atlantic Ridge 49.45 N 28.24 W
	LmH	B	58.5	H = 01 44 25.8 h = normal MB=5.4 MS=4.6
	LmV	B	59.3	D = 25.41 Az = 71.9 (NEIC)
				PV A 1.5s 85.5nm M = 5.1
				LmH B 12.5 2.8 μm 5.0
				LmV B 14 1.7 μm 4.9
5.	+eP	A	03 47 47.5	North Atlantic Ridge 25.90 N 45.00 W
				H = 03 38 58.1 h = normal MB = 5.1
				D = 49.42 Az = 44.4 (NEIC)
				PV A 1.4s 27.9nm M = 5.1
5.	eP	AB	05 53 08.5	Mediterranean Sea 35.81 N 21.84 E
	Pm	A	53 18.5	H = 05 49 17.5 h = normal MB=5.3 MS=5.1
	eS	B	56 16	D = 16.57 Az = 336.7 (NEIC)
	LmH	B	59.5	PmV A 2.0s 274.0nm M = 5.0
	LmV	B	06 00.9	LmH B 14 18.2 μm 5.5
				LmV B 16 18.8 μm 5.6
5.	eP	A	12 40 25.5	Rumania 45.56 N 26.59 E
				H = 12 37 47.2 h = 131 km MB = 4.4
				D = 11.23 Az = 302.3 (NEIC)
				PV A 1.6s 30.2nm M = 4.7
5.	eP	A	13 50 44	Sea of Okhotsk 48.16 N 146.82 E
				H = 13 39 50.4 h = 430 km MB = 5.1
				D = 74.32 Az = 332.2 (NEIC)
				PV A 1.0s 15.8nm M = 4.6
5.	ePKIKP	A	14 14 15	North Island, New Zealand
	ePKP2	AB	15 06	39.00 S 175.23 E
	ei	A	15 09.5	H = 13 54 29.1 h = 150 km MB = 6.2
	e	A	15 49	D = 163.59 Az = 320.5 (NEIC)
	ePP	AB	18 52	PKPV A 3.0s 237.0nm

Day	Phase		h m s	Remarks
cont. 5.	ePPS	B	14 32 31	PPV A 2.5s 246.0nm M = 5.8
	eSS	B	39 00	PPV B 20 4.72 /um 6.2
	ePSS	C	40 20	LmH B 23 5.1 /um
	e	B	43 50	LmV B 24 3.1 /um
	LmH	B	15 14.2	
	LmV	B	20.0	
5.	ePKHKP	A	21 47 51	<u>South of Fiji Islands</u> 21.32 S 175.20 E
				H = 21 29 12.3 h = 620 km MB = 5.2
				D = 148.10 Az = 340.1 (NEIC)
6.	ePn	A	02 08 58	<u>France</u> 46.0 N 1.2 W
	eSg	A	11 58	H = 02 06 38 (BCIS)
				D = 9.75
6.	ePKP2	A	08 17 56.5	<u>Kermadec Islands Region</u> 31.52 S 179.52 E
				H = 07 58 14.6 h = 476 km MB = 5.2 (NEIC)
				D = 158.88
6.	eP	A	15 08 29	<u>South of Honshu, Japan</u> 33.15 N 140.71 E
				H = 14 55 52.8 h = 61 km MB = 4.7
				D = 85.25 Az = 330.3 (NEIC)
				traces
6.	eP	A	15 16 00	<u>South of Alaska</u> 53.34 N 162.36 W
	e	A	16 13.5	H = 15 04 11.9 h = normal MB = 4.9
				D = 76.26 Az = 3.9 (NEIC)
6.	ePKIKP	AB	16 12 09	<u>New Hebrides Islands</u> 14.67 S 166.38 E
	ePP	AB	15 01	H = 15 52 41.9 h = 36 km MB = 6.1
	e(PcSPKP)	A	24 03	D = 138.71 Az = 335.7 (NEIC)
	eSS	B	33 12	PKPV A 2.0s 55.5nm
	LmV	B	17 14.6	PKPV B 7 0.87 /um
	LmH	B	15.4	PPV A 2.0 77.0nm M = 5.8
				PPV B 7 0.92 /um 6.2
				LmH B 19 2.1 /um 5.9
				LmV B 20 2.4 /um 5.9

Day	Phase		h m s	Remarks
6.	ePKIKP1	A	22 34 56	New Hebrides Islands 15.50 S 167.49 E
	ePKIKP2	A	35 05	H = 22 15 49.9 h = 123 km MB = 5.3
				D = 139.90 Az = 336.2 (NEIC)
				PKIKP1V A traces
				PKIKP2V A 1.2s 32.5nm
7.	ePKP	A	01 51 47.5	Loyalty Islands Region 21.88 S 170.37 E
	epPKP	A	52 06	H = 01 32 13.1 h = 66 km MB = 4.7
				D = 146.81 Az = 335.1 (NEIC)
				PKPV A 1.2s 16.3nm
7.	eP	A	12 26 04.5	Ethiopia 5.27 N 36.85 E
	e	A	26 19	H = 12 17 12.6 h = 34 km MB = 4.9
	eS	B	33 16	D = 49.92 Az = 339.2 (NEIC)
	eSS	C	36 56	PV A 1.7s 27.2nm M = 5.0
	LmH	B	48.9	LmH B 15.5 1.5 μm 5.1
	LmV	B	53.3	LmV B 15 1.1 μm 5.1
7.	ePn	A	13 08(39)	
	ePg	A	09 00	
	eSn	A	09 33	
	eSg	A	10 04	D c. 5.0
7.	e	A	14 43 52	traces
7.	eX	A	23 46 02.5	XV A 1.8s 33.8nm
8.	eP	A	04 45 37.5	Arabian Sea 13.59 N 57.35 E
	e	A	45 39	H = 04 36 26.4 h = normal MB = 5.0
				D = 52.25 Az = 324.8 (NEIC)
8.	e	A	10 04 27	Rockburst Upper Silesia, Poland
	eSg	A	04 38.5	(WAR)
8.	ePKP2	A	10 43 01	Fiji Islands Region 20.15 S 179.01 W
				H = 10 24 25.3 h = 653 km MB = 4.7 (NEIC)
				D = 148.43
				PKP2V A 1.6s 41.2nm

Day	Phase		h m s	Remarks
8.	ePKHKP	A	21 29 51.5	South of Fiji Islands 24.36 S 177.26 W
	ePKP2	A	30 02.5	H = 21 10 11.2 h = 154 km MB = 4.7 (NEIC)
	e	A	31 45	D = 157.87
				PKHKPV A 1.2s 10.2nm
9.				In the intervall $09^d 05^h 17^m - 11^h 15^m$ the short-period seismographs (type A) had been out of operation.
9.	LmH	B	06 18.3	Jan Mayen Islands Region
	LmV	B	19.3	71.27 N 7.57 W
				H = 06 02 49.1 h = 33 km MB = 4.3 (NEIC)
				D = 22.48
				LmH B 14s 0.4/um M = 4.0
				LmV B 14 0.2/um 3.9
9.	+iP	A	12 08 20.5	Southern Alaska 60.31 N 146.00 W
	e	A	08 24.5	H = 11 57 21.0 h = 18 km MB = 5.1
				D = 67.93 Az = 15.2 (NEIC)
				PV A 1.0s 37.4nm M = 5.5
9.	eP	A	13 13 52	Jan Mayen Island Region 71.4 N 7.7 W
				H = 13 08 54.3 h = 33 km MB = 4.1 (NEIC)
				D = 22.48
				PV A 1.7s 18.2nm M = 4.3
9.	e	A	13 30 24	
9.	eP	A	15 00 36.5	Jan Mayen Island Region 71.33 N 8.03 W
				H = 14 55 38.6 h = normal MB = 4.4
				D = 22.60 Az = 146.2 (NEIC)
				PV A 1.6s 30.2nm M = 4.5
9.	e	A	16 26 09	Queen Elizabeth Islands 39.5 N 73.7 E
				H = 16 17 55.2 h = 33 km MB = 4.9 (NEIC)
				D = 43.99

Day	Phase		h m s	Remarks
9.	eP	A	17 19 50	Andreanof Islands, Aleutian Is.
				51.41 N 178.21 W
				H = 17 07 55.5 h = 52 km MB = 5.1
				D = 77.99 Az = 353.6 (NEIC)
				PV A 1.2s 16.3nm M = 4.9
10.	eP1	A	03 27 36.5	Southern Greece 37.76 N 21.34 E
	+iP2	A	27 45	H = 03 24 11.7 h = 41 km MB = 5.0 (NEIC)
	LmH	B	34.3	D = 14.63
	LmV	B	34.3	P1V A 1.0s 9.8nm M = 4.3
				P2V A 1.1 48.4nm 4.9
10.	ePKS	A	11 55 11.5	Solomon Islands 11.10 S 162.28 E
	ePKS	B	55 16	H = 11 32 27.4 h = 32 km MB=5.6
	eSS	C	12 12 00	MS = 5.8 (NEIC)
	eSSS	B	17 00	D = 133.81 Az = 334.4
	LmV	B	55.5	PKSV A 1.6s 35.7nm
	LmH	B	55.6	PKSV B 12 0.65 μm
				LmH B 19.5 2.6 μm M = 5.9
				LmV B 19 3.0 μm 6.0
10.	eP	A	14 32 41.5	Jan Mayen Island Region 71.57 N 6.87 W
				H = 14 28 43.5 h = normal MB = 4.4
				D = 22.61 Az = 148.3 (NEIC)
10.	eP	A	17 24 41.5	Alaska Peninsula 55.22 N 159.94 W
				H = 17 13 08.7 h = 58 km MB = 4.2
				D = 74.27 Az = 5.6 (NEIC)
10.	e	A	20 22 19	Costa Rica 8.97 N 84.03 W
				H = 20 09 27.2 h = normal MB = 4.8
				D = 86.70 Az = 39.4 (NEIC)
				traces
11.	eP	AB	02 24 23	Fox Islands, Aleutian Is.
	eS	C	34 24	52.10 N 169.61 W
	LmH	B	03 09.5	H = 02 12 27.5 h = 30 km MB=5.4 MS=5.2

Day	Phase		h m s	Remarks
cont. 11.	LmV	B	03 11.0	D = 77.63 Az = 359.2 (NEIC)
				PV A 1.5s 75.4nm M = 5.5
				LmH B 16.5 1.0/um 5.2
				LmV B 16 0.9/um 5.2
12.	eP	A	03 30 08	Bouvet Island Region 54.41 S 5.36 E
	LmV	B	04 13.5	H = 03 16 02.8 h = 33 km MB = 5.7 (NEIC)
	LmH	B	13.6	D = 104.81
				PV A traces
				LmH B 20s 2.8/um M = 5.8
				LmV B 18 2.3/um 5.8
12.	e	A	11 12 58	
	e	A	13 20	
	e	A	13 47	
12.	eP	A	23 47 25	Hindu Kush Region 35.99 N 70.58 E
	epP	A	47 51.5	H = 23 39 26.0 h = 126 km MB = 5.2
	ePP	A	49 16	D = 44.16 Az = 308.5 (NEIC)
	e	A	49 25	h = 122 km
13.	eP	A	01 12 29	Andreanof Islands, Aleutian Is.
	epP	A	12 46	51.77 N 177.00 W
	e	A	13 07.5	H = 01 00 37.6 h = 61 km MB = 5.4
	e	A	13 15	D = 77.70 Az = 354.4 (NEIC)
				h = 63 km
13.	eP	A	06 16 46	Zambia 16.81 S 28.37 E
				H = 06 05 42.4 h = 33 km MB = 5.0
				D = 68.76 Az = 348.6 (NEIC)
				PV A 1.2s 14.2nm M = 4.9
13.	eP	A	11 23 18	Southern Sumatra 2.70 S 101.27 E
				H = 11 10 20.4 h = 105 km MB = 5.2
				D = 91.84 Az = 320.4 (NEIC)

Day	Phase		h m s	Remarks
13.	e	A	14 23 18.5	Western Pakistan 25.58 N 63.90 E H = 14 14 41.1 h = 50 km MB = 5.0 D = 46.93 Az = 316.4 (NEIC) traces
13.	e	A	17 58 35	
	e(Sg)	A	59 03	
13.	eP	A	22 10 12	Rat Islands, Aleutian Is. 51.77 N 176.26 E H = 21 58 22.0 h = 52 km MB = 5.0 D = 77.13 Az = 350.0 (NEIC) PV A 1.2s 16.3nm M = 4.9
13.	eP	A	22 14 46	Kamchatka 59.71 N 163.16 E H = 22 03 50.9 h = normal MB = 4.8 D = 67.61 Az = 340.8 (NEIC) PV A 1.6s 27.5nm M = 5.1
14.	eP	A	09 07 18.5	North Atlantic Ridge 14.77 N 45.09 W H = 08 57 28.5 h = normal MB = 5.0 D = 57.76 Az = 39.0 (NEIC)
15.	eP	A	02 04 19.5	Sakhalin Island 45.96 N 142.67 E H = 01 53 14.6 h = 354 km MB = 5.1 D = 74.88 Az = 330.2 (NEIC) h = 360 km PV A 1.4s 41.9nm M = 5.0
	epP	A	05 39	
15.	ePKP	A	03 40 18	New Hebrides Islands 20.93 S 169.77 E H = 03 20 47.3 h = 82 km MB = 5.1 D = 145.72 Az = 335.1 (NEIC)
	e(sPKP)	A	40 53	
15.	-iP	AB	09 15 07.3	Bonin Islands Region 27.08 N 140.10 E H = 09 02 58.3 h = 477 km MB = 5.8 D = 90.23 Az = 330.1 (NEIC) PV A 1.4s 298.0nm M = 6.0 PV B 2 0.55 µm 6.1 PPV A 1.8 270.0nm 6.1
	ePP	AB	18 45	

Day	Phase		h m s	Remarks
15.	eP	A	09 26 20.5	Bonin Islands Region 27.09 N 140.03 E H = 09 14 10.7 h = 478 km MB = 5.1 D = 90.19 Az = 330.1 (NEIC)
15.	+iP	A	11 32 00.3	Bonin Islands Region 27.11 N 140.03 E H = 11 19 49.9 h = 469 km MB = 4.9 D = 90.18 Az = 330.1 (NEIC) PV A 1.0s 19.7nm M = 5.0
15.	eP epP	A A	13 05 11 05 15	Southern Sinkiang Prov., China 40.43 N 91.07 E H = 12 55 44.8 h = 13 km MB = 5.1 D = 54.05 Az = 309.4 (NEIC) PV A 1.5s 32.7nm M = 5.1 pPV A 1.4 32.6nm
15.	epP	A	14 51 39	Southern Sinkiang Prov., China 40.39 N 91.06 E H = 14 42 07.8 h = 13 km MB = 4.7 D = 54.07 Az = 309.4 (NEIC)
15.	epP	A	17 16 47.5	South of Honshu, Japan 33.37 N 140.82 E H = 17 04 05.8 h = 58 km MB = 4.9 D = 85.10 Az = 330.3 (NEIC)
15.	eP	A	23 36 34.5	Nicobar Islands Region 8.9 N 94.0 E H = 23 24 31 h = 42 km D = 78.39 Az = 320 (ISC)
16.	eP	A	10 09 16.5	Fox Islands, Aleutian Is. 54.12 N 165.54 W H = 09 57 38.6 h = 81 km MB = 5.3 D = 75.58 Az = 1.9 (NEIC)
16.	ePKIKP	A	18 27 58	New Ireland Region 4.38 S 153.22 E H = 18 09 03.3 h = 50 km MB = 5.3 D = 123.78 Az = 331.6 (NEIC)

Day	Phase		h m s	Remarks
16.	eP	A	21 40 12.5	Eastern Kashmir 33.21 N 75.72 E
	epP	A	40 19	H = 21 31 26.0 h = 42 km MB = 5.1
				D = 49.19 Az = 310.8 (NEIC)
				h = 27 km
				PV A 2.0s 77.0nm M = 5.4
16.	eP	A	22 49 22.5	Mediterranean Sea 35.14 N 22.56 E
	LmH	C	57.0	H = 22 45 16.7 h = 28 km MB = 4.5
				D = 17.41 Az = 336.2 (NEIC)
17.	ePn	A	05 30 15	D c. 5
	eSg	A	31 45	
17.	ePKP	A	09 22 25.5	Fiji Islands Region 17.54 S 178.72 W
				H = 09 03 42.9 h = 531 km MB = 4.8
				D = 145.95 Az = 348.2 (NEIC)
17.	-iPKP	AB	10 03 43	Tonga Islands 15.13 S 175.01 W
				H = 09 44 36.8 h = 251 km MB = 5.7
				D = 144.17 Az = 352.8 (NEIC)
				PKPV A 1.4s 477.0nm
17.	eP	A	14 10 52	Jan Mayen Island Region 71.18 N 8.22 W
				H = 14 05 56.7 h = normal MB = 4.3
				D = 22.52 Az = 145.7 (NEIC)
18.	eP diff	C	09 43 50	New Hebrides Islands 6.87 S 149.99 E
	ePKIKP	AB	47 10	H = 09 28 14.1 h = 43 km MB=6.3 MS=6.8
	eiX	A	47 18.5	D = 124.34 Az = 329.2 (NEIC)
	iPP	B	48 58	PKIKPV A 1.3s 65.5nm
	eiSKSP	B	58 54	XV A 2.3 598.0nm
	eiSS	C	10 05 40	PPV B 11 6.8/um M = 6.8
	eiSSS	C	10 10	LmH B 20 21.9/um 6.8
	LmV	B	41.5	LmV B 22 30.7/um 6.9
	LmH	B	42.5	

Day	Phase		h m s	Remarks
18.	ePn	A	19 09 51	Yugoslavia 42.8 N 19.2 E
	e	A	12 03	H = 19 07 36 (BCIS)
				D = 9.36 Az = 329 (ISC)
				PV A 0.8s 13.5nm
18.	ePg	A	21 53 54	France 45.31 N 6.75 E
	eSg	A	55 28.5	H = 21 51 56.4 h = normal
				D = 6.26 Az = 29.7 (NEIC)
19.	e	A	01 02 29.5	Kodiak Island Region 57.05 N 153.94 W
				H = 00 50 57.3 h = 32 km MB = 4.9
				D = 72.01 Az = 9.6 (NEIC)
19.	eP	A	14 13 48	South of Honshu, Japan 33.14 N 140.69 E
				H = 14 01 15.0 h = 60 km MB = 4.7
				D = 85.25 Az = 330.3 (NEIC)
19.	eP	A	15 18 16	Afghanistan 32.71 N 68.35 E
	e	A	19 15	H = 15 10 02.1 h = 30 km MB = 5.0
				D = 44.83 Az = 311.0 (NEIC)
19.	ePKHKP	A	23 44 09	Tonga Islands Region 23.44 S 175.64 W
				H = 23 24 14.8 h = normal MB=5.0 MS=5.0
				D = 152.26 Az = 350.0 (NEIC)
				PKHKPV A 2.0s 42.7nm
20.	ePKP2	A	08 52 52	South of Kermadec Islands
				33.34 S 178.32 W
				H = 08 32 10.9 h = 33 km MB = 4.9 (NEIC)
				D = 161.23
20.	eP	A	10 26 17.5	Near East Coast of Honshu, Japan
				34.95 N 140.84 E
				H = 10 13 51.3 h = 43 km MB = 5.0
				D = 83.74 Az = 330.3 (NEIC)
				PV A 1.8s 40.5nm M = 5.2

Day	Phase		h m s	Remarks
20.	ePg	A	10 40 47	France 46.0 N 1.4 W
	eSg	A	42 55	H = 10 37 29 (BCIS)
				D = 9.87
20.	eP	A	12 55 21	West Pakistan 29.49 N 68.60 E
	Pm	A	55 27	H = 12 46 44.0 h = 15 km MB = 5.0
	LmH	B	13 19.3	D = 47.14 Az = 313.3 (NEIC)
	LmV	B	23.3	PmV A 1.8s 74.4nm M = 5.4
				LmH B 15 4.3 μm 5.5
				LmV B 12 3.6 μm 5.6
20.	LmH	B	13 08.8	West Pakistan 29.28 N 68.57 E
	LmV	B	08.8	H = 12 34 19.6 h = 17 km MB = 5.3 (NEIC)
				D = 47.27
				LmH B 13s 3.9 μm M = 5.6
				LmV B 13 2.9 μm 5.5
20.	eP	A	16 42 28.5	Near East Coast of Honshu, Japan
				34.88 N 140.95 E
				H = 16 29 59.1 h = 20 km MB = 5.0
				D = 83.85 Az = 330.3 (NEIC)
20.	e	A	17 07 58	Near East Coast of Honshu, Japan
				34.87 N 140.88 E
				H = 16 55.19.8 h = 33 km MB = 4.9
				D = 83.83 Az = 330.3 (NEIC)
				traces
21.	eP	A	04 51 17	West Pakistan 29.27 N 68.56 E
	LmH	C	05 16.0	H = 04 42 43.4 h = 20 km MB = 5.3
	LmV	B	19.0	D = 47.27 Az = 313.4 (NEIC)
				PV A 1.0s 15.7nm M = 5.0
21.	eP	A	05 33 59.5	West Pakistan 29.31 N 68.49 E
				H = 05 25 23.4 h = 11 km MB = 5.2 (NEIC)
				D = 47.20

Day	Phase		h m s	Remarks
21.	eP	A	08 28 35	Honshu, Japan 36.11 N 139.77 E
				H = 08 16 18.8 h = 61 km MB = 5.0
				D = 82.31 Az = 329.7 (NEIC)
21.	ePKP	A	10 57 09	Tonga Islands 15.94 S 174.03 W
				H = 10 37 39.8 h = 87 km MB = 4.3
				D = 145.09 Az = 353.7 (NEIC)
				PKPV A 1.7s 42.4nm
21.	e	A	16 18 11	Kodiak Island Region 58.30 N 151.76 W
				H = 16 06 45.5 h = 29 km MB = 4.6
				D = 70.56 Az = 11.1 (NEIC)
				traces
21.	ePKP	A	20 40 12	Tonga Islands 15.86 S 174.06 W
				H = 20 20 49.3 h = 129 km MB = 5.5
				D = 145.00 Az = 353.7 (NEIC)
				PKPV A 1.0s 51.1nm
22.	eP1	AB	00 51 04.5	Off Coast of Jalisco, Mexico
	eP2	A	51 11	18.60 N 104.97 W
	ePP	AB	54 43	H = 00 37 58.0 h = normal MB=5.6 MS=6.1
	eSKS	B	01 01 42	D = 91.47 Az = 34.7 (NEIC)
	eS	C	02 10	P1V A 2.0s 64.1nm M = 5.6
	ePS	B	03 16	P2V A 2.0 103.0nm 5.8
	eSS	B	08 20	PPV A 2.4 194.5nm 6.1
	LmH	B	33.2	LmH B 17.5 15.1/um 6.5
	LmV	B	33.2	LmV B 18 15.3/um 6.5
22.	ePKIKP	A	13 55 39.5	Fiji Islands Region 21.66 S 179.04 W
	-iPKHKP	AB	55 44.8	H = 13 36 57.8 h = 570 km MB = 5.2
	ePKP2	A	55 51.5	D = 149.88 Az = 346.4 (NEIC)
	epPKHKP	A	58 14.5	PKHKPV A 1.6s 126.0nm
22.	ePKIKP	A	24 04 30	New Britain Region 5.97 S 149.69 E
	ePS	C	16 15	H = 23 45 36.7 h = 72 km MB = 5.2
	eSS	C	23 50	D = 123.42 Az = 329.4 (NEIC)
	LmH	B	59.2	LmH B 20s 3.6/um
	LmV	B	59.3	LmV B 20 3.7/um

Day	Phase		h m s	Remarks
23.	ePKHKP	A	05 08 47.5	Santa Cruz Islands 12.09 S 166.50 E
	e	A	08 50.5	H = 04 49 45.7 h = 97 km MB = 5.8
	ePKIKP	AB	08 58	D = 136.41 Az = 336.9 (NEIC)
	ePP	AB	11 42	PKIKPV A 3.0s 657.0nm
	e(ScPPKP)	A	20 28	PKIKPV B 11.0 2.06 μm
	e	A	20 49	PPV A 1.6 77.0nm M = 5.6
	eSS	B	29 32	PPV B 12.0 2.7 μm 6.3
	eSSS	B	34 46	LmH B 26 6.8 μm
	LmH	B	06 02.4	LmV B 24 5.7 μm
	LmV	B	03.0	
23.	-eP	A	11 51 03.5	Crete 34.31 N 25.06 E
				H = 11 46 42.6 h = normal MB = 4.6
				D = 19.05 Az = 333.0 (NEIC)
				PV A 0.9s 38.9nm M = 4.6
23.	eSn	A	14 00 25.5	Yugoslavia 43.35 N 20.6 E
	eSb	A	01 21	H = 13 56 21.3 h = 0 km (ISC)
				D = 9.6
24.	eP	A	03 29 02	Southern Sinkiang Prov., China
	(LmH)	B	51.8	40.96 N 82.18 E
				H = 03 20 20.2 h = normal MB = 5.1
				D = 48.40 Az = 306.6 (NEIC)
				PV A 1.4s 18.6nm M = 4.9
24.	ePn	A	07 59 43	Czechoslovakia 50°47.4' N 14°31.5' E
	eiSg	A	08 00 08	Explosion yield 13 t (PRU)
				D = 1.84
25.	+iP	A	18 43 50.7	Near East Coast of Kamchatka
				54.57 N 161.60 E
				H = 18 32 27.4 h = 31 km MB=5.3 MS=4.3
				D = 72.19 Az = 340.5 (NEIC)
				PV A 1.2s 75.1nm M = 5.6
26.	eP	A	07 54 03.5	Mediterranean Sea 35.82 N 22.10 E
	e	A	54 29	H = 07 50 12.8 h = 58 km MB = 4.7
	LmH	C	08 00.0	D = 16.64 Az = 336.1 (NEIC)

Day	Phase		h m s	Remarks
26.	eP	A	13 18 18	Carlsberg Ridge 1.18 S 67.37 E
				H = 13 07 07.1 h = normal MB = 5.1
				D = 69.98 Az = 325.9 (NEIC)
26.	ePKP	A	13 24 59	Tonga Islands 17.52 S 173.04 W
				H = 13 05 16.8 h = normal MB=5.2 MS=5.2
				D = 146.75 Az = 354.6 (NEIC)
27.	eP	A	04 16 20	Kurile Islands 50.40 N 156.85 E
				H = 04 04 41.6 h = 47 km MB = 5.2
				D = 75.06 Az = 337.9 (NEIC)
				traces
27.	eP	A	13 22 42.5	Northern Celebes 0.12 S 123.93 E
	ePP	AB	27 04	H = 13 08 43.7 h = 55 km MB = 6.0
	ePPS	B	37 05	D = 104.08 Az = 322.6 (NEIC)
	ePKKP	A	38 32	PV A 1.3s 19.7nm M = 5.8
	LmH	B	14 15.0	PPV A 2.0 128.0nm 6.2
	LmV	B	15.0	LmH B 17.5 0.8 µm 5.3
				LmV B 18 0.8 µm 5.3
27.	eSg	A	19 54 19	Poland 50.27 N 18.92 E
				H = 19 51 44.8 M = 2.8 (WAR)
				D = 4.65
27.	ePKP	A	20 57 41.5	Tonga Islands Region 17.51 S 172.85 W
				H = 20 38 00.3 h = normal MB = 5.1
				D = 146.75 Az = 354.8 (NEIC)
				PKPV A 1.3s 48.0nm
28.	ePKIKP	A	08 36 20	Fiji Islands Region 19.74 S 177.99 W
	ePKHKP	A	36 24.5	H = 08 17 25.4 h = 413 km MB = 4.6
	ePKP	A	36 28	D = 148.23 Az = 348.4 (NEIC)
				PKIKPV A traces
				PKHKPV A 1.7s 18.2nm
28.	ePg	A	12 30 18.5	Yugoslavia 45.5 N 14.9 E
	eSg	A	31 33	H = 12 28 36
				D = 5.57 Az = 338 (ISC)

Day	Phase		h m s	Remarks
28.	+eiPKP	A	14 37 40.5	Tonga Islands 16.51 S 173.76 W
	+eX	A	37 58.5	H = 14 18 03.1 h = normal MB = 5.1
				D = 145.68 Az = 353.9 (NEIC)
				PKPV A 1.4s 53.5nm
				XV A 1.3 43.7nm
28.	eP	A	17 13 36	Hokkaido, Japan Region 41.85 N 141.11 E
				H = 17 01 46.3 h = 93 km MB = 5.1
				D = 77.88 Az = 329.8 (NEIC)
				traces
28.	+iPKHKP	A	17 54 02	Tonga Islands 19.28 S 175.93 W
	e(PKP2)	A	54 04	H = 17 34 42.1 h = 236 km MB = 5.0
	epPKHKP	A	55 03.5	D = 148.13 Az = 350.9 (NEIC)
	e	A	55 55	h = 246 km
				PKHKPV A 1.8s 74.3nm
28.	+iPKP	AB	18 02 44	New Hebrides Islands 19.85 S 169.03 E
	eX	A	02 55	H = 17 43 14.6 h = 72 km MB = 5.7
	LmV	B	19 09.7	D = 144.45 Az = 335.1 (NEIC)
	LmH	B	09.8	PKPV A 1.3s 11.6nm
				PKPV B 5 2.64 µm
				XV A 1.4 32.1nm
				LmH B 21 1.6 µm
				LmV B 22 1.6 µm
28.	eP	A	20 39 09	Ionian Sea 38.17 N 19.55 E
				H = 20 35 55.0 h = normal MB = 4.3
				D = 13.69 Az = 338.2 (NEIC)
28.	iPg	A	21 55 48.3	
	iSg	A	56 11	
29.	eP	A	04 40 28	Northwestern Kashmir 35.82 N 73.29 E
				H = 04 32 06.6 h = 53 km MB = 5.0 (NEIC)
				D = 45.99
				traces

Day	Phase	h m s	Remarks
29.	eP A	06 05 47.5	Ionian Sea 38.18 N 19.87 E
	e A	09 01	H = 06 02 32.4 h = normal MB = 4.3
			D = 13.78 Az = 337.4 (NEIC)
			PV A 0.9s 13.6nm M = 4.8
29.	e A	10 27 12.5	
29.	ePKIKP A	14 46 02	Tonga Islands 21.89 S 174.94 W
	ePKHKP A	46 06.5	H = 14 26 15.7 h = normal MB=5.1 MS=4.9
	e A	46 20	D = 150.85 Az = 351.4 (NEIC)
			PKIKP A traces
			PKHKP A 1.7s 66.6nm
29.	eP A	21 36 12	Southwestern Ryukyu Islands
	LmH B	22 19.0	23.95 N 123.44 E
	LmV B	19.0	H = 21 23 39.6 h = 53 km MB = 4.9
			D = 84.56 Az = 323.6 (NEIC)
30.	eP A	02 39 42	Spain 37.00 N 3.62 W
	e A	40 10	H = 02 36 11.6 h = 634 km MB = 4.0
			D = 17.47 Az = 33.9 (NEIC)
30.	LmH B	16 12.0	Probably: Off Coast of Southern Chile
	LmV B	14.0	45.68 S 76.64 W
			H = 15 03 23.9 h = 33 km MB = 5.0 (NEIC)
			D = 122.43
30.	+eP1 AB	21 14 11.5	Near Coast of Michoacan, Mexico
	+iP2 A	14 14.5	18.48 N 103.00 W
	+iP3 AB	14 22.5	H = 21 01 12.5 h = 43 km MB=6.2 MS=7.5
	e B	14 44	D = 90.50 Az = 35.4 (NEIC)
	ePP B	18 10	P1V A 1.7s 90.9nm M = 5.9
	eS B	25 10	P2V A 1.8 213.0nm 6.2
	ePS B	26 50	P3V A 2.4 1720.0nm 7.0
	ePKKP3 A	31 42	P3V B 7.5 13.2µm 7.4
	eSS B	31 50	PPV B 16 20.2µm 7.3
	eP'P'3 A	40 03.5	LmH B 17.5 195.0µm 7.6

Day	Phase		h m s	Remarks
cont. 30.	e(P'PKS)	A	21 43 09	LmV B 17s 234.0 µm M = 7.7
	LmV	B	56.9	
	LmH	B	57.5	
31.	+eiP	AB	21 07 54	Bonin Islands Region 28.21 N 139.23 E
	epP	AB	09 43	H = 20 55 53.1 h = 498 km MB = 6.0
	+ePP	AB	11 35	D = 88.88 Az = 329.7 (NEIC)
	e	B	14 08	PV A 2.0s 513.0nm M = 6.0
	eiS	B	17 50	PPV A 1.5 262.0nm 6.1
	eisS	B	21 12	SH B 16.5 16.3 µm 6.7
	iSS	B	24 00	LmH B 15.5 12.1 µm
	eSKPP'	A	36 31.5	LmV B 14 8.3 µm
	LmH	B	48.3	
	LmV	B	54.5	

Day	Phase		h m s	Remarks
1.	eX	A	03 01 03	XV A 1.4s 25.6nm
1.	eP	AB	05 27 42	Jujuy Province, Argentina
	epP	AB	28 38	66.20 W 22.66 S
	ePP	AB	31 45	H = 05 14 20.6 h = 229 km MB = 6.1
	epPP	AB	32 35	D = 99.85 Az = 39.2 (NEIC)
	esPP	B	33 00	h = 232 km
	iSKS	B	38 00	PV A 1.5s 95.5nm M = 6.0
	iS	B	38 54	PV B 5 1.2/um 6.6
	ePKKP	A	44 00	PPV A 2.0 145.0nm 5.7
	e	A	44 26	SH B 14 3.4/um 6.3
	e	C	47 12	
	e	C	51 00	
	eP'P'	A	52 21	
1.	eiPKP	AB	07 46 59	Tonga Islands 17.70 S 175.16 W
	epPKP	A	48 04	H = 07 27 44.8 h = 232 km MB = 5.3
				D = 146.69 Az = 352.1 (NEIC)
				h = 260 km
				PKPV A 2.6s 701.0nm
				PKPV B 7 3.48/um
1.	ePKP2	A	08 32 07	South of Fiji Islands 25.35 S 176.13 W
	epPKP2	A	32 19	H = 08 11 53.4 h = 50 km MB = 4.9
				D = 154.05 Az = 348.7 (NEIC)
1.	eP	A	17 35 51	Rat Islands, Aleutian Is.
	e	A	35 56.5	51.79 N 176.26 E
	e(S)	C	46 10	H = 17 24 00.9 h = 51 km MB = 5.3
	LmH	B	18 09.6	D = 77.11 Az = 350.0 (NEIC)
	LmV	B	13.5	PV A 1.4s 39.6/um M = 5.2
				LmH B 20 1.2/um
				LmV B 20 0.9/um
2.	ePKP	A	08 42 11	Fiji Islands Region 17.55 S 178.87 W
	e	A	42 50.5	H = 08 23 36.6 h = 609 km MB = 4.5
				D = 145.93 Az = 348.1 (NEIC)

Day	Phase		h m s	Remarks
3.	eSg	A	05 28 18.5	Yugoslavia 43.8 N 16.2 E H = 05 24 25 (BCIS) D = 7.5
3.	e	A	08 55 29	
4.	ePKHKP	A	10 49 34.5	South of Fiji Islands 26 S 179 E H = 10 29 37 MB = 4.3 (NORSAR)
4.	eX	A	15 45 24	XV A 1.6s 22.0nm
4.	e(P)	A	19 34 04	Afghanistan 31.67 N 67.26 E H = 19 25 48.4 h = 40 km MB = 3.9 D = 44.83 Az = 311.8 (NEIC) traces
5.	eP epP	A A	04 42 29.5 42 49.5	Kurile Islands 43.92 N 147.47 E H = 04 30 33.1 h = 53 km MB = 5.3 D = 78.30 Az = 333.1 (NEIC) h = 78 km PV A 1.4s 32.6nm M = 5.1
5.	e(S)	A	20 37 44	Yugoslavia 41.4 N 20.6 E H = 20 32 41 (BCIS) D = 11.2
5.	e	A	23 39 27	Northern Sinkiang Prov., China 42.90 N 82.07 E H = 23 30 48.6 h = normal MB = 4.5 D = 47.19 Az = 305.1 (NEIC)
6.	eiP	A	05 42 23.5	Ryukyu Islands 27.87 N 127.69 E H = 05 30 02.0 h = 81 km MB = 5.7 D = 83.63 Az = 324.9 (NEIC) PV A 1.5s 68.0nm M = 5.4
6.	eP1 iP2 eiP3	AC AB A	10 47 53.5 47 59.5 48 00	Szechwan Prov., China 31.40 N 100.58 E H = 10 37 10.1 h = normal MB=6.1 MS=7.4 D = 65.82 Az = 315.8 (NEIC)

Day	Phase		h m s	Remarks
cont. 6.	+iP4	AB	10 48 10	P1V A 1.2s 30.5nm M = 5.3
	i(PcP)	B	48 41	P4V A 2.0 700.0nm 6.4
	i	B	49 32	P4V B 7 7.0 µm 6.9
	iPP	B	50 36	SH B 16 32.8 µm 7.1
	iPPP	B	52 08	LmH B 23.5 1350.0 µm 8.1
	+iS	CN	56 44	LmV B 14 362.0 µm 7.8
	+iS	CE	56 47	P'P' traces
	iSS	B	11 01 00	
	iSSS	B	04 11	
	eSSSS	B	05 53	
	e(PKKP)	A	08 38	
	LmH	B	13.3	
	P'P'	A	16 42	
	LmV	B	19.1	
7.	ePKP	A	02 27 57	Tonga Islands Region 17.58 S 172.69 W H = 02 08 16.9 h = normal MB=5.0 MS=5.0 D = 146.83 Az = 355.0 (NEIC) PKPV A 1.9s 106.0nm
7.	ePn	A	04 02 44	Yugoslavia 42.99 N 17.76 E
	ePg	A	03 28	H = 04 00 35.4 h = normal
	eSn	A	04 19.5	D = 8.73 Az = 333.3 (NEIC)
	eSg	A	04 48	
7.	e	A	05 34 04.5	Western Iran 32.05 N 49.43 E
	e	A	36 30	H = 05 27 20.0 h = 51 km MB = 5.2 D = 33.45 Az = 314.9 (NEIC)
7.	e(PP)	A	08 20 23	South Sandwich Islands Region 55.91 S 27.84 W H = 08 01 13.7 h = normal MB = 5.7 (NEIC) D = 111.13 PPV A 1.8s 27.0nm M = 5.7
7.	eP	A	09 52 18.5	Jan Mayen Island Region 71.56 N 6.91 W H = 09 47 18.7 h = normal MB = 4.4 D = 22.61 Az = 148.3 (NEIC) PV A 0.8s 15.4nm M = 4.5

Day	Phase		h m s	Remarks
7.	-iP1	AB	16 17 08	Szechwan Prov., China 31.46 N 100.29 E
	eiP2	AB	17 11.5	H = 16 06 25.0 h = normal MB=5.8 MS=5.5
	ePP	A	19 36	D = 65.60 Az = 315.7 (NEIC)
	eS	C	25 52	P1V A 1.7s 218.0nm M = 6.0
	eSS	C	30 10	P2V A 1.8 358.0nm 6.2
	eSSS	C	33 06	LmH B 15.5 3.7/um 5.7
	LmV	B	47.8	LmV B 13 3.7/um 5.8
	LmH	B	47.9	
8.	eP	A	02 46 41.5	Kurile Islands 50.46 N 156.43 E
				H = 02 35 09.0 h = 80 km MB = 4.5
				D = 74.89 Az = 337.7 (NEIC)
				PV A 1.3s 17.5nm M = 4.8
8.	eP	A	03 40 40	Greece 38.28 N 20.34 E
	e	A	40 47	H = 03 37 23.1 h = 32 km MB = 4.2 (NEIC)
				D = 13.83
8.	eP	A	09 29 04	Greece 38.26 N 20.26 E
				H = 09 25 46.3 h = 53 km MB = 4.1
				D = 13.83 Az = 336.4 (NEIC)
8.	ePKIKP	A	09 57 37.5	Off Coast of Southern Chile
	LmH	B	10 54.4	44.96 S 77.85 W
	LmV	B	54.8	H = 09 38 43.7 h = normal MB = 5.4
				D = 122.61 Az = 49.1 (NEIC)
				LmH B 18s 1.3/um M = 5.6
				LmV B 16 1.1/um 5.6
8.	ePKIKP	A	10 27 57.5	Southeast Indian Rise 45.48 S 96.29 E
	e	A	28 02.5	H = 10 09 08.3 h = normal MB=5.9 MS=6.1
	LmH	B	11 26.8	D = 120.42 Az = 312.7 (NEIC)
	LmV	B	28.4	LmH B 18s 2.5/um M = 5.9
				LmV B 19 3.3/um 6.0
8.	iPg	A	11 45 41.8	Czechoslovakia 50.56 N 14.00 E
	iSg	A	46 02.4	Explosion 14 to (PRU)
				D = 1.52

Day	Phase		h m s	Remarks
8.	eP	A	12 00 34.5	Greece 38.43 N 20.01 E H = 11 57 19.7 h = 17 km MB = 4.5 D = 13.59 Az = 336.7 (NEIC) PV A 1.0s 11.8nm M = 4.8
8.	eP	A	16 47 45	Greece 38.38 N 20.13 E H = 16 44 31.5 h = normal MB = 4.6 D = 13.67 Az = 336.5 (NEIC) PV A 0.9s 15.6nm M = 4.8
8.	e	A	19 04 25	Probably: South of Fiji Islands 25 S 178 W H = 18 44 19 (NORSAR) D = 137.9 traces
8.	eP	A	19 15 56	Ascension Island Region 10.38 S 13.01 W
	LmH	B	44.4	H = 19 05 21.9 h = normal MB=5.3 MS=5.3
	LmV	B	45.4	D = 64.44 Az = 17.1 (NEIC) PV A 2.3s 177.0nm M = 5.8 LmH B 16 0.6/um 4.8 LmV B 16 0.8/um 5.1
9.	ePP	A	05 21 06.5	Celebes 2.71 S 119.84 E H = 05 02 43.6 h = 33 km MB = 5.2 (NEIC) D = 103.60
9.	e	A	10 05 52	Colombia 4.63 N 76.03 W H = 09 53 18.5 h = 84 km MB = 5.0 D = 84.96 Az = 39.7 (NEIC) traces
9.	eP	A	11 58 39	North of Ascension Island 0.50 S 16.16 W H = 11 49 00.6 h = normal MB = 4.6 D = 56.18 Az = 20.9 (NEIC)
10.	LmH	C	04 26.0	Taiwan 24.27 N 121.75 E
	LmV	B	27.0	H = 03 32 28.0 h = 56 km MB = 4.8 D = 83.38 Az = 323.0 (NEIC) LmH C 22s 0.7/um LmV B 19 0.5/um

Day	Phase		h m s	Remarks
10.	eP1	A	05 24 49	South of Honshu, Japan 33.20 N 140.65 E
	eP2	A	24 51	H = 05 12 17.7 h = 62 km MB = 5.3
	e	A	25 04	D = 85.18 Az = 330.2 (NEIC)
				P2V A traces
				P2V A 1.3s 21.8nm M = 5.0
10.	ePKHKP	A	11 07 58	Tonga Islands Region 22.29 S 174.86 W
				H = 10 48 05.4 h = normal MB = 4.8
				D = 151.25 Az = 351.4 (NEIC)
				PKHKPV A 1.8s 27.0nm
10.	eP	AB	12 06 29	Near Coast of Michoacan, Mexico
	ePP	B	10 08	18.89 N 103.55 W
	eSKS	C	17 08	H = 11 53 27.5 h = normal MB=5.4 MS=5.6
	ePS	C	18 44	D = 90.47 Az = 35.2 (NEIC)
	eSS	C	23 30	PV A 2.3s 158.3nm M = 5.9
	LmH	B	48.6	PV B 8 1.15/um 6.3
	LmV	B	48.7	LmH B 16.5 7.0/um 6.2
				LmV B 16.5 7.4/um 6.2
10.	eP	A	14 48 22	Carlsberg Ridge 2.52 N 66.42 E
				H = 14 37 35.1 h = normal MB = 5.3
				D = 66.41 Az = 325.4 (NEIC)
				PV A 2.6s 173.2nm M = 5.7
10.	eP	A	17 07 16	Kurile Islands 49.90 N 156.08 E
	e	A	09 45	H = 16 55 33.9 h = normal MB=5.5 MS=5.2
	LmH	B	42.7	D = 75.34 Az = 337.5 (NEIC)
	LmV	B	43.4	PV A 1.5s 65.4nm M = 5.4
				LmH B 20.5 1.8/um 5.4
				LmV B 20 1.0/um 5.2
10.	eP	A	18 16 11.5	Carlsberg Ridge 2.64 N 66.33 E
				H = 18 05 23.9 h = normal MB = 4.9
				D = 66.26 Az = 325.4 (NEIC)
10.	eP	A	22 08 33.5	Ascension Island Region 6.86 S 12.27 W
				H = 21 58 21.2 h = normal MB = 4.7
				D = 60.89 Az = 17.2 (NEIC)

Day	Phase		h m s	Remarks
10.	eP	A	22 19 40	Carlsberg Ridge 2.55 N 56.29 E H = 22 08 52.8 h = normal MB = 5.0 D = 66.29 Az = 325.4 (NEIC)
10.	ePKHKP ePKP2	A A	22 52 33 52 43.5	South of Fiji Islands 23.37 S 179.97 E H = 22 33 42.1 h = 570 km MB = 4.7 D = 151.30 Az = 344.5 (NEIC)
11.	eX	A	09 07 30.5	Crete 35.70 N 23.92 E H = 09 03 22.7 h = 31 km MB = 4.2 D = 17.40 Az = 333.0 (NEIC) XV A 1.3s 15.3nm M = 4.0
11.	ePKIKP	A	10 51 29.5	East New Guinea Region 6.55 S 147.44 E H = 10 32 40.1 h = 78 km MB = 5.1 D = 122.76 Az = 328.2 (NEIC) PKIKPV A 1.1s 10.1nm
11.	epP	A	11 31 57	Mindanao, Philippine Islands 6.66 N 126.84 E H = 11 17 58.3 h = 71 km MB = 5.1 D = 100.41 Az = 324.2 (NEIC)
11.	ePKHKP	A	15 04 35	Kermadec Islands 29.82 S 177.56 W H = 14 44 29.8 h = 63 km MB = 4.6 D = 158.09 Az = 344.2 (NEIC) traces
11.	ePKHKP ePKP2	A A	15 10 15.5 10 23.5	Fiji Islands Region 21.05 S 178.01 W H = 14 51 28.3 h = 571 km MB = 4.5 D = 149.51 Az = 347.9 (NEIC) PKHKPV A 1.1s 10.1nm
12.	ePKP2	A	04 55 31	Fiji Islands Region 20.60 S 176.91 W H = 04 36 25.1 h = 411 km MB = 4.6 D = 149.27 Az = 349.4 (NEIC)

Day	Phase		h m s	Remarks
13.	-iPKP	AB	15 41 33	Fiji Islands Region 17.52 S 178.50 W
				H = 15 22 55.1 h = 541 km MB = 5.5
				D = 145.97 Az = 348.5 (NEIC)
				PKPV A 1.6s 412.0nm
13.	ePKP	A	18 21(11)	New Hebrides Islands 20.48 S 169.66 E
	epPKP	A	21 47.5	H = 18 01 49.8 h = 143 km MB=5.2 (NEIC)
	e	A	22 00	D = 145.27
				PKPV A traces
13.	eP	A	20 05 48	Andreanof Islands, Aleutian Is.
				51.25 N 179.22 W
				H = 19 53 53.5 h = 46 km MB=5.4 MS=4.7
				D = 78.07 Az = 353.0 (NEIC)
				PV A 1.1s 40.3nm M = 5.4
14.	+iP	AB	01 01 48.3	Taiwan Region 22.29 N 121.55 E
	i	A	01 53.8	H = 00 49 16.2 h = 38 km MB=5.9 MS=5.1
	LmV	B	44.8	D = 84.85 Az = 323.1 (NEIC)
	LmH	B	44.9	PV A 1.2s 167.0nm M = 6.1
				LmH B 16.5 6.2 μm 6.1
				LmV B 16 9.7 μm 6.3
14.	eP	A	07 23 16.5	South of Alaska 53.77 N 158.90 W
				H = 07 11 32.0 h = 32 km MB = 4.9
				D = 75.65 Az = 6.2 (NEIC)
				PV A 1.0s 19.7nm M = 5.1
14.	+iPKHKP	A	09 46 47.6	Fiji Islands Region 20.05 S 178.31 W
				H = 09 28 06.4 h = 600 km MB = 5.8
				D = 148.48 Az = 347.9 (NEIC)
				PKHKPV A 1.0s 31.0nm
14.	e	A	09 54 26	
	e	A	55 32	
14.	ePKIKP	A	16 31 12	Kermadec Islands 30.39 S 177.51 W
	e	A	31 15.5	H = 16 11 14.9 h = 29 km MB=5.7 MS=6.3

Day	Phase		h m s	Remarks
cont. 14.	ePKP2	A	16 31 46	D = 158.64 Az = 343.9 (NEIC)
	e	A	31 57	LmH C 17s 7.5 µm M = 6.5
	eSS	C	55 30	LmV C 16 14.3 µm 6.9
	LmH	C	17 53.8	
	LmV	C	58.3	
14.	eP	A	21 57 44	<u>Honshu, Japan</u> 39.03 N 141.47 E
	epP	A	58 11	H = 21 45 42.4 h = 110 km MB = 5.1
	esP	A	58 23	D = 80.46 Az = 330.3 (NEIC)
	ePP	A	22 00 43	h = 109 km
				PV A traces
				pPV A 1.5s 37.7nm
14.	eP	A	22 48 31	<u>Near East Coast of Honshu, Japan</u>
				37.04 N 141.52 E
				H = 22 36 14.6 h = 56 km MB = 5.1
				D = 82.20 Az = 330.5 (NEIC)
				PV A 1.6s 24.7nm M = 5.0
15.	LmH	C	01 07.0	Probably <u>Near Coast of Nicaragua</u>
	LmV	C	08.0	11.53 N 86.47 W
				H = 00 15 53.7 h = 33 km MB = 5.1 (NEIC)
				D = 86.25
15.	eP	A	05 44 28.5	<u>Southern Sumatra</u> 4.34 S 103.09 E
				H = 05 31 19.5 h = 107 km MB = 5.4
				D = 94.26 Az = 320.3 (NEIC)
				PV A 1.5s 27.7nm M = 5.4
15.	eP	A	07 40 00.5	<u>Near East Coast of Kamchatka</u>
	LmH	C	08 27.0	56.06 N 163.55 E
				H = 07 28 44.5 h = 45 km MB=4.8 MS=4.1
				D = 71.14 Az = 341.5 (NEIC)
				LmH C 26s 0.9 µm M = 4.9
15.	eP	A	07 56 18	<u>South of Panama</u> 5.99 N 82.37 W
				H = 07 43 27.4 h = normal MB=5.3 MS=5.3
				D = 87.95 Az = 39.5 (NEIC)

Day	Phase		h m s	Remarks
15.	e	A	11 27 42	Probably Kermadec Islands 30.38 S 177.30 W H = 11 06 59.9 h = 25 km MB = 4.7 (NEIC) D = 158.68
15.	eP	A	21 28 18	Greece 38.32 N 20.28 E
	e	A	28 21	H = 21 25 03.7 h = normal MB = 4.3 D = 13.77 Az = 336.3 (NEIC) PV A 0.7s 9.6nm M = 4.7
15.	eP	A	22 29 31	Kurile Islands 47.00 N 153.82 E H = 22 17 37.0 h = normal MB = 4.8 D = 77.42 Az = 336.4 (NEIC) PV A 1.0s 13.8nm M = 4.9
16.	ePKP	A	05 10 33.5	Tonga Islands 15.31 S 173.32 W
	LmH	C	06 07.2	H = 04 51 01.4 h = 50 km MB=5.8 MS=5.2
	LmV	B	17.4	D = 144.53 Az = 354.6 (NEIC) PKPV A 1.6s 44.0nm LmH C 25 0.9/um LmV B 20 0.7/um
16.	+iP	AB	05 10 46.6	Eastern Kazakh 49.84 N 78.23 E
	ePn	A	12 19	H = 05 02 57.7 h = 0 km MB = 5.6 (NEIC) Underground explosion M = 5.8 (UPP) D = 41.30 PV A 0.8s 138.5nm M = 5.7
16.	e	A	08 29 34	Czechoslovakia 50.81 N 15.07 E
	e	A	29 36	explosion 11.7 t
	e	A	29 49	H = 08 29.3 (PRU) D = 2.2
16.	ePKP2	A	12 14 49	Kermadec Islands 30.22 S 177.19 E
	e	A	14 55	H = 11 54 18.8 h = 42 km MB = 5.1 (NEIC) D = 158.56 PKP2V A 1.2s 16.3nm

Day	Phase		h m s	Remarks
16.	ePn	A	20 52 07	Central Italy 42.9 N 12.6 E
				H = 20 50 10 (BCIS)
				D = 7.82
17.	eP	A	16 13 35.5	Leeward Islands 16.95 N 61.42 W
	LmH	C	35.4	H = 16 02 45.5 h = 34 km MB=5.5 MS=4.5
				D = 66.34 Az = 41.7 (NEIC)
				LmH C 25s 0.6 um M = 4.7
17.	+eiP	A	19 26 55.5	Kurile Islands 45.21 N 148.57 E
				H = 19 15 11.4 h = 113 km MB = 5.3
				D = 77.50 Az = 333.6 (NEIC)
				PV A 1.2s 104.0nm M = 5.5
17.	e(P)	A	19 48 21.5	Greenland Sea 73.2 N 5.2 E
				H = 19 43 11 h = 0 km
				D = 22.77 Az = 169 (ISC)
17.	LmH	C	23 23.0	Southwest of Africa 52.80 S 9.88 E
				H = 22 33 37.4 h = normal MB = 4.9 (NEIC)
				D = 103.0
				LmH C 25s 0.7 um M = 4.9
18.	eP	A	04 01 52	South Atlantic Ridge 36.69 S 17.20 W
	ePP	A	05 21.5	H = 03 48 52.9 h = normal MB=5.6 MS=5.4
	eS	C	12 40	D = 90.60 Az = 17.9 (NEIC)
	ePS	C	13 50	PV A 2.4s 117.5nm M = 5.8
	eiSS	B	18 24	PPV A 2.2 82.0nm 5.8
	LmH	B	38.0	LmH B 20 0.8 um 5.1
	LmV	B	38.2	
18.	ePKP2	A	07 52 22	Kermadec Islands 30.21 S 177.46 W
				H = 07 31 50.8 h = 20 km MB = 4.7 (NEIC)
				D = 158.48
				PKP2V A 2.0s 42.7nm
18.	ePKIKP	A	18 29 21	Kermadec Islands 30.23 S 177.61 W
	eiPKP2	AB	29 56	H = 18 09 27.2 h = 49 km MB=5.5 MS=5.7

Day	Phase		h m s	Remarks
cont. 18.	ePP	B	18 33 34	D = 158.47 Az = 343.9 (NEIC)
	eSS	C	53 30	PKIKPV A 1.5s 20.1nm
	LmH	B	19 44.0	PKP2V A 1.8 94.5nm
	LmV	B	56.6	PPV A 10 0.75 µm M = 5.7
				LmH B 20 1.1 µm 5.6
				LmV B 18 1.5 µm 5.9
18.	eP	A	20 28 18	Central Italy 42.7 N 12.5 E
				H = 20 26 20 (BCIS)
				D = 8.02
18.	ePKP2	A	20 33 47	Kermadec Islands Region 30.4 S 176.9 W
				H = 20 13 07 h = 49 km
				D = 158.77 Az = 345 (ISC)
				PKP2V A 1.2s 10.2nm
18.	LmH	B	22 03.8	Kirgiz-Sinkiang Border Region
	LmV	B	08.0	48.83 N 74.14 E
				H = 21 39 02.3 h = normal MB = 4.9 (NEIC)
				D = 39.4
				LmH B 15s 0.9 µm M = 4.7
19.	ePP	C	09 00 28	Prince Edward Islands Region
	eSKS	B	07 15	45.54 S 35.13 E
	eiPS	B	09 24	H = 08 42 52.1 h = 33 km MB = 5.5 (NEIC)
	eSS	B	14 35	D = 97.95
	LmV	B	41.3	LmH B 18s 4.4 µm M = 6.0
	LmH	B	41.4	LmV B 17 4.4 µm 6.0
19.	e	A	09 01 44.5	Probably Turkey 38 N 29 E
				H = 08 56 27 (NORSAR)
19.	e	A	17 47 15	
19.	+eP	A	18 14 20	Turkey 40.19 N 33.85 E
	LmH	B	21.2	H = 18 10 00.5 h = 22 km MB = 4.5

Day	Phase		h m s	Remarks
cont. 19.	LmV	B	18 23.2	D = 18.72 Az = 311.3 (NEIC)
				PV A 1.2s 28.4nm M = 4.3
				LmH B 20 2.8 µm 4.6
				LmV B 16 2.2 µm 4.7
19.	e(Sg)	A	20 33 40	<u>France</u> 46.0 N 1.5 W
				H = 20 27 50 (BCIS)
				D = 9.92
20.	LmH	C	02 53.0	<u>Balleny Islands Region</u> 62.50 S 166.27 E
				H = 01 14 40.7 h = normal MB = 5.7 (NEIC)
				D = 65.5
				LmH C 20s 0.6 µm M = 4.8
20.	eP	A	05 59 31	<u>Crete</u> 34.41 N 23.84 E
	e	A	59 38	H = 05 55 14.4 h = 22 km MB = 4.5
	eiX	A	59 51	D = 18.52 Az = 334.9 (NEIC)
	LmV	B	06 08.3	PV A 0.8s 7.7nm M = 3.9
	LmH	B	08.4	XV A 1.0 25.6nm
				LmH B 16 0.5 µm 3.9
				LmV B 15 0.7 µm 4.2
20.	+iP	A	07 51 50.5	<u>Gulf of Alaska</u> 58.31 N 149.81 W
	eiX	AB	51 55.5	H = 07 40 34.8 h = 12 km MB=5.5 MS=5.1
	e	A	52 05.5	D = 70.35 Az = 12.4 (NEIC)
	eS	C	08 01 00	PV A 1.3s 43.6nm M = 5.4
	eSS	C	05.5	XV A 1.5 95.5nm
	LmH	B	29.8	LmH B 15 0.8 µm 5.1
	LmV	B	32.7	LmV B 15 1.0 µm 5.2
20.	e	A	11 43 13	<u>Crimea Region</u> 45.02 N 36.41 E
				H = 11 38 48.8 h = 17 km MB = 4.2
				D = 17.54 Az = 297.6 (NEIC)
20.	e	A	11 47 11	
	e	A	47 55	
	e	A	48 36	

Day	Phase		h m s	Remarks
20.	eP	A	14 24 28	South of Honshu, Japan 32.43 N 137.42 E H = 14 12 37.8 h = 401 km MB = 4.4 D = 84.46 Az = 328.7 (NEIC) traces
20.	ePKP2	A	21 27 36.5	Kermadec Islands 30.36 S 177.13 W H = 21 07 07.2 h = 33 km MB = 4.6 (NEIC) D = 158.71
20.	ePKP2	A	22 26 10	Balleny Islands Region
	LmH	C	23 43.0	62.59 S 167.45 E
	LmV	B	55.0	H = 22 05 19.6 h = 33 km MB = 5.4 (NEIC) D = 162.25 PKP2V traces LmH C 21s 0.6/um M = 5.3 LmV B 17 0.5/um 5.4
21.	eP	A	01 18 13	Greece 38.25 N 20.15 E H = 01 14 59.9 h = normal MB = 4.1 D = 13.80 Az = 336.7 (NEIC)
21.	eP	A	08 33 56	Szechwan Province, China 31.67 N 100.06 E H = 08 23 15.2 h = normal MB = 4.9 D = 65.32 Az = 315.5 (NEIC) PV A (1.6)s 16.5nm M = 4.9
21.	eP	AB	14 58 33.5	Southern California 34.07 N 119.04 W
	e	A	15 01 44.5	H = 14 45 57.3 h = 8 km MB=5.7 MS=5.2
	ePP	AB	01 49.5	D = 85.09 Az = 29.0 (NEIC)
	eS	B	09 10	PV A 2.4s 283.0nm M = 6.1
	eSS	C	14 45	PPV A 2.6 155.0nm 6.1
	eSSS	C	18.8	LmH B 17.5 3.9/um 5.9
	LmH	B	37.6	LmV B 14 4.0/um 6.0
	LmV	B	39.7	

Day	Phase		h m s	Remarks
21.	eSg	A	20 11 06	Poland 50.37 N 18.82 E H = 20 08 36.0 M = 3.2 (WAR) D = 4.6
22.	-eP	A	00 42 56	Guatemala 14.48 N 91.64 W
	eS	C	53 32	H = 00 30 20.3 h = 107 km MB = 5.2
	LmH	B	01 19.0	D = 87.12 Az = 38.4 (NEIC)
	LmV	B	19.0	PV A 1.3s 35.0nm M = 5.2 LmH B 20 0.7/um LmV B 20 0.7/um
22.	e(Pg)	A	06 49 47	France 48.17 N 6.58 E
	iSg	A	50 42.5	H = 06 48 33.0 h = normal D = 4.12 Az = 51.1 (NEIC) (Pg)V A 0.8s 32.7nm
22.	ePKIKP	A	16 41 12.5	New Britain Region 4.62 S 151.83 E H = 16 22 32.2 h = 154 km MB = 5.3 D = 123.32 Az = 330.8 (NEIC)
22.	ePKP2	A	20 24 23.5	Kermadec Islands 30.34 S 177.06 W H = 20 03 58.2 h = 75 km MB = 4.7 (NEIC) D = 158.70
23.	eP	A	04 39 24	Ecuador 2.01 S 78.22 W
	e	A	39 30.5	H = 04 26 23.3 h = 67 km MB = 5.7
	LmH	B	05 15.5	D = 91.49 Az = 39.6 (NEIC)
	LmV	B	15.5	PV A 2.0s 68.4nm M = 5.7 LmH B 21 1.4/um
23.	ePKIKP	A	11 27 24	New Britain Region 7.00 S 150.02 E H = 11 08 23.9 h = 21 km MB = 5.2 D = 124.47 Az = 329.2 (NEIC)
23.	eP	A	19 49 28	Costa Rica 9.85 N 83.49 W
	e	A	49 32	H = 19 36 48.7 h = normal MB=4.7 MS=4.2 D = 85.69 Az = 39.5 (NEIC)

Day	Phase		h m s	Remarks
23.	ePKIKP	A	21 16 20.5	South of Fiji Islands 23.90 S 179.97 W
	ePKHKP	A	16 27	H = 20 57 31.0 h = 518 km MB = 5.0
	ePKP2	A	16 39	D = 151.82 Az = 344.3 (NEIC)
				PKIKPV A traces
				PKHKPV A 1.2s 20.3nm
				PKP2V A 1.2 26.4nm
24.	eP	A	00 10 00	Southern Iran 28.58 N 52.62 E
				H = 00 02 40.1 h = 27 km MB = 5.2
				D = 37.85 Az = 317.1 (NEIC)
24.	ePKP	A	01 25 46.5	Tonga Islands 15.41 S 173.28 W
	LmV	C	02 32.5	H = 01 06 11.3 h = normal MB=5.1 MS=5.3
	LmH	C	34.8	D = 144.63 Az = 354.6 (NEIC)
				PKPV A 1.3s 17.5nm
				LmH C 18 0.8/um M = 5.5
				LmV C 18 1.6/um 5.8
24.	e(PKP)	A	04 16 25	Tonga Islands 15.17 S 173.44 W
				H = 03 56 43.5 h = normal MB=5.2 MS=5.2
				D = 144.38 Az = 354.5 (NEIC)
				PKPV A 1.4s 18.6nm
24.	+eiPKP	AB	07 57 54	New Hebrides Islands 19.19 S 168.74 E
	ePP	A	08 01 12	H = 07 38 27.0 h = 59 km MB = 6.0
	LmV	B	09 05.9	D = 143.74 Az = 335.3 (NEIC)
	LmH	B	06.0	PKPV A 1.6s 280.0nm
				PPV A 1.8 77.7nm M = 5.6
				LmH B 20 1.2/um
				LmV B 20 1.2/um
24.	ePKP	A	20 39 04.5	New Hebrides Islands 19.16 S 168.72 E
				H = 20 19 35.2 h = 45 km MB = 4.5
				D = 143.71 Az = 335.3 (NEIC)
24.	eP1	A	22 05 38	West Pakistan 26.73 N 66.41 E
	eP2	A	05 40	H = 21 57 04.3 h = 54 km MB = 4.2
				D = 47.68 Az = 315.3 (NEIC)
				P2V A 1.3s 21.8nm M = 5.0

Day	Phase		h m s	Remarks
24.	eP	A	23 58 14	Crete 34.96 N 24.48 E
				H = 23 54 03.4 h = 50 km MB = 4.1
				D = 18.26 Az = 333.1 (NEIC)
25.	ePKIKP	A	05 54 40	Scotia Sea 61.04 S 37.88 W
	ePP	AB	55 55	H = 05 35 55.4 h = 33 km MB=6.4 MS=6.2
	ePPP	B	58 25	D = 118.19 Az = 33.3 (NEIC)
	eSKS	B	06 01 20	PKIKPV A 1.6s 44.0nm
	eSKKS	B	02 52	PPV A 2.0 295.0nm M = 6.7
	ePKKP	A	05 07	PPV B 5 1.65 /um 7.0
	iSP	C	05 25	LmH B 21.5 5.0 /um 6.1
	ePPS	B	06 53	LmV B 16.5 4.1 /um 6.1
	eSS	C	11 58	
	eiPSPS	C	12 34	
	eSSS	C	16 48	
	LmH	B	37.9	
	LmV	B	44.4	
25.	e	A	07 35 08	
	i	A	35 20	
25.	eP	AB	10 44 38	Southern Sumatra 1.70 S 99.68 E
	e	A	44 59	H = 10 31 39.5 h = normal MB=5.4 MS=5.9
	iS	B	55 28	D = 90.07 Az = 320.5 (NEIC)
	ePS_1	C	56 40	PV A 1.6s 33.0nm M = 5.4
	ePS_2	C	57 00	LmH B 17 2.0 /um 5.6
	eSS	C	11 01 18	LmV B 21 2.6 /um 5.7
	eSSS	C	05 15	
	eSSSS	C	08 24	
	LmV	B	31.4	
	LmH	B	32.3	
25.	eP	A	21 41 26	Southern Sumatra 1.69 S 99.66 E
				H = 21 28 26.3 h = normal MB = 4.9
				D = 90.05 Az = 320.5 (NEIC)
				traces

Day	Phase		h m s	Remarks
26.	eP	A	00 56 20.5	**Costa Rica** 9.61 N 84.23 W H = 00 43 39.9 h = normal MB = 4.7 MS=4.0 D = 86.34 Az = 39.4 (NEIC) PV A 1.5s 32.6nm M = 5.3
26.	eP	A	08 28 39	**Kurile Islands** 49.41 N 155.62 E H = 08 16 57.6 h = 46 km MB = 5.0 D = 75.67 Az = 337.3 (NEIC)
26.	eP	A	11 07(12)	**Philippine Islands Region** 5.70 N 127.04 E H = 10 53 27.6 h = 74 km MB = 5.2 D = 101.30 Az = 324.1 (NEIC) LmH B 18s 0.7/um LmV B 18 0.7/um
	LmH	B	56.0	
	LmV	B	56.0	
26.	eP	A	22 10 28	**Southern Sumatra** 1.76 S 99.66 E H = 21 57 30.0 h = normal MB=5.2 MS=5.3 D = 90.10 Az = 320.5 (NEIC) PV A 1.4s 18.6nm M = 5.2 LmH B 19 0.9/um 5.2 LmV B 18 0.9/um 5.3
	eS	B	21 18	
	eSSSS	C	34 00	
	LmV	B	57.5	
	LmH	B	58.2	
26.	eP	A	22 26 06.5	**Greece-Albania Border Region** 39.97 N 20.10 E H = 22 23 11.69 h = normal MB = 4.6 D = 12.22 Az = 333.7 PV A 0.7s 17.2nm M = 5.3
27.	LmH	B	10 10.0	**Philippine Islands Region** 6.16 N 127.06 E H = 09 08 57.0 h = 68 km MB = 5.2 (NEIC) D = 101.0
	LmV	B	15.0	
27.	e	A	15 02 46	**Southern Sumatra** 1.75 S 99.66 E H = 14 49 37.0 h = normal MB = 5.0 D = 90.10 Az = 320.5 (NEIC)

Day	Phase	h m s	Remarks
27.	eP A	17 14 15	<u>Turkey</u> 38.85 N 29.85 E
	LmH B	20.2	H = 17 10 10.6 h = 33 km MB = 4.2
	LmV B	22.0	D = 17.46 Az = 318.4 (NEIC)
			PV A 2.3s 85.4nm M = 4.5
			LmH B 15.5 0.9/um 4.1
			LmV B 13 0.7/um 4.2
27.	eP A	18 17 44	<u>Fox Islands, Aleutian Is.</u>
			52.46 N 168.70 W
			H = 18 05 53.2 h = 47 km MB = 4.7
			D = 77.27 Az = 359.8 (NEIC)
			PV A 1.3s 17.5nm M = 4.9
28.	eP1 AB	06 49 29	<u>Kurile Islands</u> 50.49 N 156.58 E
	eiP2 AB	49 37.5	H = 06 37 49.5 h = 27 km MB=6.3 MS=7.2
	iPP C	52 24	D = 74.92 Az = 337.8 (NEIC)
	iPPP B	54 00	P1V A 2.1s 436.0nm M = 6.1
	eS B	58 56	P2V A 2.5 2920.0nm 6.9
	eP'P' A	07 16 47	P2V B 14 38.7/um 7.2
	LmH B	27.7	P'P'V A 1.6 22.0nm
	LmV B	27.9	LmH B 18.5 342.0/um 7.7
			LmV B 18 283.0/um 7.7
28.	+eP A	07 02 21	<u>Kurile Islands</u> 50.11 N 156.93 E
			H = 06 50 39.7 h = normal MB = 5.3
			D = 75.35 Az = 338.0 (NEIC)
			PV A 1.4s 48.8nm M = 5.3
28.	eP A	07 07 18	<u>Kurile Islands</u> 50.08 N 156.94 E
			H = 06 55 38.7 h = 48 km MB = 5.5
			D = 75.38 Az = 338.0 (NEIC)
			PV A 1.8s 98.0nm M = 5.4
28.	e(P) A	08 21 06.5	<u>Mindanao, Philippine Islands</u>
			9.55 N 126.13 E
			H = 08 07 34.8 h = 81 km MB = 5.3
			D = 97.67 Az = 324.2 (NEIC)

Day	Phase		h m s	Remarks
28.	eP	A	11 44 22.5	Kurile Islands 50.12 N 156.92 E
	e	A	44 45	H = 11 32 42.7 h = 48 km MB = 5.2
	LmV	B	12 22.2	D = 75.34 Az = 338.0 (NEIC)
	LmH	B	22.3	PV A 1.7s 54.5nm M = 5.2
				LmH B 16 1.5/um 5.4
				LmV B 17 1.1/um 5.3
28.	e(PKP2)	A	14 13 45	South of Fiji Islands 26 S 179 E
				H = 13 53 44 (NORSAR)
				PKP2V A 1.0s 15.7nm
28.	ePKP	A	21 04 53.5	Fiji Islands Region 18.25 S 178.49 W
				H = 20 46 11.5 h = 540 km MB = 4.4 (NEIC)
				D = 146.69
				PKPV traces

Day	Phase		h m s	Remarks
1.	eP	A	00 53 23	Central Mid-Atlantic Ridge 1.35 N 29.52 W H = 00 43 16.3 h = normal MB = 4.6 D = 60.16 Az = 28.9 (NEIC)
1.	+eiP epP	A A	02 30 47 30 59	Kurile Islands Region 49.88 N 157.30 E H = 02 19 03.1 h = 30 km MB = 5.5 D = 75.65 Az = 338.3 (NEIC) h = 50 km PV A 1.5s 78.0nm M = 5.5 pPV A 1.5 45.2nm
1.	ePKHKP	A	03 40 13	Tonga Islands Region 23.28 S 175.42 W H = 03 20 19.9 h = normal MB = 4.9 D = 152.13 Az = 350.4 (NEIC)
1.	ePKP	A	14 25 45.5	Tonga Islands 17.40 S 173.09 W H = 14 06 04.5 h = normal MB = 4.4 D = 146.62 Az = 354.6 (NEIC)
2.	eP	A	03 02 34.5	Greenland Sea 73.38 N 7.7 E H = 02 57 26.5 h = 0 km MB = 4.3 D = 22.88 Az = 174 (ISC) PV A 1.6s 16.5nm M = 4.2
2.	e	A	19 33 51.5	Turkey 39.38 N 27.69 E H = 19 30 01.9 h = 55 km MB = 3.6 D = 15.96 Az = 320.1 (NEIC) traces
3.	eP LmH LmV	A B B	01 56 22.5 02 33.6 39.0	Ryukyu Islands 24.99 N 128.20 E H = 01 43 41.8 h = normal MB = 5.3 D = 86.25 Az = 325.2 (NEIC) PV A 1.8s 23.6nm M = 5.1 LmH B 17 1.0/um 5.3 LmV B 17 0.9/um 5.2

Day	Phase		h m s	Remarks
3.	+iP	A	02 53 46.5	Kurile Islands 50.44 N 156.28 E
				H = 02 42 09.2 h = 59 km MB = 5.5
				D = 74.88 Az = 337.6 (NEIC)
				PV A 1.2s 24.4nm M = 5.0
3.	iPg	A	05 36 49.0	Vogtland, German Democratic Republic
	eSg	A	36 56.0	50.3 N 12.5 E
				H = 05 36 37 (BCIS)
				D = 0.67
3.	iPg	A	07 41 06.5	Vogtland, German Democratic Republic
	eSg	A	41.14.5	(MOX)
3.	iPg	A	10 53 54	Vogtland, German Democratic Republic
	eSg	A	54 01	50.34 N 12.39 E
				H = 10 53 42.2 h = 6 km
				D = 0.58 Az = 302.0 (NEIC)
4.	iPg	A	06 24 49.3	Vogtland, German Democratic Republic
	i	A	24 54.3	(MOX)
	iSg	A	24 56.5	
4.	iPg	A	07 14 33.2	Vogtland, German Democratic Republic
	eSg	A	14 40.8	(MOX)
4.	eP	A	14 22 32.5	Ascension Island Region 7.3 S 13.3 W
	e	A	22 45	H = 14 12 15.7 h = normal MB = 5.1 (NEIC)
				D = 61.63
				PV A 1.4s 23.2nm M = 5.1
4.	eiP	AB	18 09 05.5	Near East Coast of Kamchatka
	ePP	B	12 00	54.83 N 161.60 E
	ePPP	C	13 45	H = 17 57 43.5 h = 32 km MB=6.1 MS=5.8
	eS	B	18 24	(NEIC)
	ePS	B	19 09	D = 71.94
	eSS	C	23 33	PV A 1.4s 1210.0nm M = 6.7
	e	C	28.0	LmH B 19 6.4 μm 5.9
	LmH	B	39.9	LmV B 15 5.3 μm 6.0
	LmV	B	49.2	

Day	Phase		h m s	Remarks
5.	eP	A	15 43 48.5	Greece-Albania Border Region 39.4 N 20.6 E H = 15 40 45 (BCIS) D = 12.93 traces
5.	ePn	A	20 07 02	Ruhr Region, Federal Republic of Germany
	e	A	07 07.5	51.5 N 7.3 E
	ePg	A	07 12	Rockburst
	eSg	A	07 48	H = 20 06 15 (BCIS) D = 2.82
5.	eP	A	23 34 07.5	Mindanao, Philippine Islands
	epP	A	34 29	6.15 N 126.85 E H = 23 20 26.0 h = 84 km MB = 5.3 (NEIC) D = 100.82 h = 83 km PV A 1.4s 23.2nm M = 4.6
5.	eP	A	24 05 42	United Arab Republic 27.69 N 33.64 E
	e	A	05 46	H = 23 59 46.6 h = 25 km MB = 4.5 (NEIC)
	e	A	05 51	D = 28.39 PV A 1.4s 18.6nm M = 4.6
6.	eP	A	04 11 19.5	Luzon, Philippine Islands 18.11 N 120.71 E H = 03 58 40.2 h = 98 km MB = 5.1 (NEIC) D = 87.70
6.	iPg	A	11 23 49.6	Kraslice, Czechoslovakia
	iSg	A	24 06.3	(PRU)
6.	eP	A	22 32 22	Baja California 27.54 N 112.45 W
	LmV	B	23 11.5	H = 22 19 32.7 h = 33 km MB = 5.1 (NEIC)
	LmH	B	12.0	D = 87.82 LmH B 14.5s 1.8/um M = 5.6 LmV B 16.5 1.7/um 5.6

Day	Phase		h m s	Remarks
7.	eP	A	03 30 21	Bonin Islands Region 27.74 N 139.86 E
	ePP	A	34 03.5	H = 03 18 13.2 h = 465 km MB = 4.9 (NEIC)
				D = 89.56
				PV A 1.4s 16.3nm M = 5.2
7.	eP	A	06 37 43.5	Albania 41.80 N 20.05 E
	e	A	37 53	H = 06 35 10.7 h = 48 km MB = 4.0
				D = 10.60 Az = 329.5 (NEIC)
				PV A 1.2s 8.1nm M = 4.7
7.	ePKP2	A	08 45 23	Kermadec Islands Region 30.97 S 179.00 W
				H = 08 25 02.9 h = 114 km MB = 5.0 (NEIC)
				D = 158.80
				PKPV A 1.0s 11.8nm
7.	ePKP2	A	13 15 39	South of Kermadec Islands
	e	A	15 43	33.49 S 179.64 W
				H = 12 54 55.6 h = 6 km MB = 5.6 (NEIC)
				D = 160.97
7.	e	A	16 28 58	
8.	e	A	03 29 15.5	
8.	ePKIKP	A	13 21 48.5	Fiji Islands Region 17.71 S 178.86 W
	ePKHKP	A	21 50.5	H = 13 03 13.5 h = 582 km MB = 4.8
				D = 146.09 Az = 348.0 (NEIC)
				PKHKPV A 1.6s 88.0nm
8.	ePKHKP	A	14 55 13	Tonga Islands Region 23.02 S 175.59 W
	ePKP2	AB	55 23.5	H = 14 35 18.4 h = normal MB=5.0 MS=5.6
	LmV	B	16 10.1	D = 151.86 Az = 350.2 (NEIC)
	LmH	B	11.0	LmH B 18.5s 1.0/um M = 5.6
				LmV B 19 0.7/um 5.5
8.	ePKP2	A	15 29 17	Tonga Region 22.4 S 175.9 W
				H = 15 09 10.9 h = 0 km MB = 4.4
				D = 151.18 Az = 350 (ISC)

Day	Phase		h m s	Remarks
8.	e(PKP2)	A	15 58 40	<u>Tonga Region</u> 22.60 S 175.58 W
				H = 15 38 30.5 h = 0 km MB = 4.6
				D = 151.45 Az = 350 (ISC)
8.	eP	A	16 22 18.5	<u>Southern Nevada</u> 37.10 N 116.03 W
				H = 16 10 00.2 h = 0 km MB = 5.4
				D = 81.24 Az = 30.7 (NEIC)
				Nuclear explosion (AEC)
				PV A 1.3s 28.4nm M = 5.2
8.	e	A	18 11 45	<u>Kermadec Islands Region</u> 32 S 176 W
				H = 17 51 22 (NORSAR)
				D = 160.7
8.	LmV	B	21 57.6	Probably <u>Off Coast of Jalisco, Mexico</u>
	LmH	B	57.7	18.49 N 105.01 W
				H = 21 02 32.1 h = normal MB = 4.6 (NEIC)
				D = 90.5
				LmH B 18s 0.3 µm M = 4.9
				LmV B 16 0.5 µm 5.1
9.	eP	AB	10 20 23	<u>Philippine Islands Region</u>
	ePP	B	24 30	6.27 N 127.34 E
	eSKS	B	30 58	H = 10 06 37.7 h = 55 km MB = 6.0
	eS	B	31 53	D = 101.01 Az = 324.2 (NEIC)
	iSP	AB	33 25	PV A 1.9s 129.0nm M = 6.2
	e(SS)	C	40 23	PV B 10 1.3 µm 6.4
	LmV	B	11 12.8	SPV A 3.0 210.0nm
	LmH	B	15.5	SPV B 8 2.5 µm
				LmH B 17.5 8.7 µm
				LmV B 19 9.7 µm
9.	LmH	B	14 48.6	Probably <u>Kurile Islands</u>
	LmV	B	49.0	49.98 N 157.07 E
				H = 13 58 41.3 h = normal MB = 4.6 (NEIC)
				D = 75.6

Day	Phase		h m s	Remarks
9.	+eiPKP	A	19 28 53	<u>Near S.E. Coast of Australia</u>
	e	A	28 58.5	34.13 S 150.27 E
				H = 19 09 12.8 h = 13 km MB = 5.5
				D = 145.80 Az = 311.5 (NEIC)
9.	e	A	21 31 00	
10.	ePKP	A	10 09 37.5	<u>Tonga Islands</u> 15.18 S 173.52 W
	e	A	09 51	H = 09 50 00.3 h = 8 km MB=5.6 MS=5.1
	LmH	C	11 06.1	D = 144.39 Az = 354.4 (NEIC)
				PKPV A 1.7s 33.4nm
				LmH C 26 0.45 µm M = 5.7
10.	LmH	C	11 06.1	
10.	ePKP	A	14 38 45.5	<u>Samoa Islands Region</u> 15.39 S 172.70 W
				H = 14 19 11.5 h = normal MB = 4.9
				D = 144.66 Az = 355.2 (NEIC)
				PKPV A 1.9s 30.3nm
10.	ePKHKP	A	18 27 50	<u>Tonga Islands Region</u> 23.06 S 175.40 W
	ePKP2	A	27 58	H = 18 07 56.2 h = normal MB=5.0 MS=5.2
	LmH	B	19 43.5	D = 151.92 Az = 350.5 (NEIC)
	LmV	B	43.5	
10.	iPg	A	21 46 57.5	<u>Vogtland, German Democratic Republic</u>
	eSg	A	47 05.5	(MOX)
10.	ePg	A	21 50 11.5	<u>Vogtland, German Democratic Republic</u>
	eiSg	A	50 19.4	(MOX)
10.	iPg	A	21 59 18.6	<u>Vogtland, German Democratic Republic</u>
	iSg	A	59 27	(MOX)
11.	eP	A	04 29(15)	<u>Western Mediterranean Sea</u> 37.0 N 1.0 E
				H = 04 25 29 (BCIS)
				D = 15.61 Az = 26 (ISC)

Day	Phase		h m s	Remarks
11.	iPg	A	06 58 28.9	Vogtland, German Democratic Republic
	iSg	A	58 36.4	50.44 N 12.43 E
				H = 06 58 15.8 h = 0 km
				D = 0.56 Az = 292 (ISC)
11.	iPg	A	09 05 34.1	Vogtland, German Democratic Republic
	eiSg	A	05 41.5	50.36 N 12.23 E
				H = 09 05 20.3 h = 0 km
				D = 0.48 Az = 307 (ISC)
11.	iPg1	A	12 17 28.5	Vogtland, German Democratic Republic
	iPg2	A	17 32.5	50.41 N 12.57 E
	eiSg1	A	17 36	H = 12 17 19.9 h = normal
	iSg2	A	17 40	D = 0.65 Az = 291.6 (NEIC)
	LmH	B	17.7	LmH B 4s 2.87/um
11.	ePg	A	12 21 28	Vogtland, German Democratic Republic
	eSg	A	21 36	(MOX)
11.	eP	A	13 49 42	Hokkaido, Japan Region 41.59 N 142.01 E
	e	A	49 53	H = 13 37 47.3 h = 73 km MB = 5.5
	e	A	50 01	D = 78.44 Az = 330.3 (NEIC)
	e	A	50 15	LmH B 18s 0.5/um
	LmH	B	14 21.8	LmV B 16 0.5/um
	LmV	B	27.4	
11.	eP	A	15 05 42	Taiwan Region 21.03 N 120.16 E
	LmH	B	43.3	H = 14 53 07.5 h = 32 km MB = 4.8
	LmV	B	47.6	D = 85.07 Az = 322.7 (NEIC)
				PV A 1.8s 27.0nm M = 5.1
				LmH B 16 0.5/um 5.0
				LmV B 16 0.4/um 4.9
11.	ePKHKP	A	15 44 43.5	Fiji Islands Region 19.98 S 178.30 W
				H = 15 26 03.9 h = 618 km MB = 4.9
				D = 148.41 Az = 347.9 (NEIC)

Day	Phase		h m s	Remarks
12.	eP	A	03 32 11.5	<u>Fox Islands, Aleutian Is.</u>
				52.61 N 168.06 W
				H = 03 20 16.8 h = 15 km MB = 4.8
				D = 77.12 Az = 0.2 (NEIC)
12.	+eP	AB	11 26 03	<u>Kurile Islands</u> 50.10 N 156.74 E
	LmH	B	12 04.0	H = 11 14 23.6 h = 49 km MB=5.7 MS=5.2
	LmV	B	04.6	D = 75.31 Az = 337.9 (NEIC)
				PV A 1.5s 105.5nm M = 5.6
				LmH B 15 3.1 /um 5.7
				LmV B 17 2.5 /um 5.6
12.	eP	A	12 52 56	<u>Talaud Islands</u> 4.18 N 126.49 E
	LmH	B	13 39.0	H = 12 39 02.0 h = 37 km MB=5.5 MS=5.4
	LmV	B	46.5	D = 102.20 Az = 323.8 (NEIC)
				PV A 1.7s 42.5nm M = 5.8
				LmH B 18 2.1 /um 5.7
				LmV B 18.5 2.7 /um 5.8
12.	e	A	13 28 28	<u>Western Iran</u> 32.05 N 49.35 E
				H = 13 21 49.1 h = 62 km MB = 4.9
				D = 33.40 Az = 315.0 (NEIC)
12.	eP	A	16 32 27	<u>Greenland Sea</u> 73.62 N 8.62 E
	LmH	B	41.8	H = 16 27 24.0 h = 33 km MB = 4.9
	LmV	B	42.3	D = 23.09 Az = 175.1 (NEIC)
				PV A 1.4s 44.2nm M = 4.8
				LmH B 15 0.5 /um 4.1
				LmV B 15 0.5 /um 4.2
12.	iPn	A	17 18 07.5	<u>Northern Italy</u> 44.1 N 12.1 E
	eSg	A	20 14	H = 17 16 35 (BCIS)
				D = 6.55
12.	eP	A	17 49 17.5	<u>Greenland Sea</u> 73.75 N 9.20 E
				H = 17 44 13.9 h = normal MB = 4.6
				D = 23.21 Az = 176.1 (NEIC)
				PV A 1.5s 20.1nm M = 4.4

Day	Phase		h m s	Remarks
12.	+iP	AB	19 50 56.5	<u>Kurile Islands</u> 50.84 N 157.10 E
	ePa	B	55 47	H = 19 39 21.0 h = 54 km MB = 6.1
	eS	B	20 00 28	D = 74.72 Az = 338.0 (NEIC)
	eSS	C	05 32	PV A 2.5s 454.0nm M = 6.0
	LmH	B	27.5	PV B 8 2.6 μm 6.2
	LmV	B	29.5	LmH B 19 11.4 μm
				LmV B 17 10.0 μm
				Pa: velocity = 8.42 km/s
12.	eP	A	20 34 33	<u>Mediterranean Sea</u> 35.91 N 21.78 E
				H = 20 30 43.9 h = 44 km MB = 4.7
				D = 16.46 Az = 336.6 (NEIC)
				PV A 1.1s 24.2nm M = 4.2
13.	ePKIKP	A	02 01 24.5	<u>Solomon Islands</u> 5.45 S 154.21 E
	epPKIKP	A	02 13	H = 01 42 43.6 h = 170 km MB = 5.5
	ePP	A	03 18.5	D = 125.18 Az = 331.8 (NEIC)
	LmH	C	55.5	h = 197 km
				LmH C 25s 0.37 μm
13.	ePg	A	03 05 57	<u>Vogtland, German Democratic Republic</u>
	eSg	A	06 06	(MOX)
13.	eP	A	06 10 35	<u>Western Persia</u> 32.0 N 49.6 E
				H = 06 03 53 (BCIS)
				D = 33.60
13.	ePKIKP	A	20 20 07.5	<u>New Britain Region</u> 6.35 S 153.07 E
				H = 20 01 08.4 h = 35 km MB = 5.3
				D = 125.43 Az = 330.9 (NEIC)
				PKIKPV A 1.2s 8.1nm
14.	e	A	03 52 34	<u>Western Persia</u> 32.0 N 50.0 E
				H = 03 45 36 (BCIS)
				D = 33.85
14.	ePKP2	A	06 28 16	<u>Balleny Islands Region</u> 62.47 S 165.27 E
	e	A	28 23	H = 06 07 30.8 h = 36 km MB = 5.5 (NEIC)

Day	Phase		h m s	Remarks
cont. 14.	LmV	B	07 53.9	D = 161.43
	LmH	B	54.6	LmH B 18s 0.8/um M = 5.5
				LmV B 19 1.1/um 5.7
14.	ePKIKP	A	11 44 39.5	New Britain Region 5.32 S 152.15 E
				H = 11 25 46.7 h = 64 km MB = 5.8
				D = 124.08 Az = 330.8 (NEIC)
				PKIKPV A 1.9s 30.3nm
15.	eP	A	07 44 58	North Atlantic Ridge 10.73 N 43.44 W
	eS	C	53 18	H = 07 34 50.5 h = normal MB = 4.5
	LmH	B	08 08.5	D = 59.93 Az = 37.1 (NEIC)
	LmV	B	09.7	PV A 1.0s 7.9nm M = 4.8
				SH C traces
				LmH B 20s 0.4/um 4.6
				LmV B 22 0.5/um 4.7
15.	ePKP	A	08 49 59	Tonga Islands 15.49 S 175.26 W
				H = 08 30 20.5 h = 33 km MB = 5.3 (NEIC)
				D = 144.50
				PKPV A 2.1s 28.8nm
15.	e(Pg)	A	09 42 33	
	e(Sg)	A	43 46	
15.	eP	A	18 11 31.5	Mindoro, Philippine Islands
				13.94 N 120.36 E
				H = 17 58 39.4 h = 113 km MB = 5.2
				D = 90.81 Az = 322.9 (NEIC)
16.	eP	AB	01 05 50.5	Molucca Passage 2.09 N 126.64 E
	ePP	AB	10 13	H = 00 51 47.0 h = 18 km MB=6.0 MS=6.3
	eiSKS	B	16 32	D = 103.96 Az = 323.5 (NEIC)
	eS	B	17 36	PV A 2.3s 73.1nm M = 6.1
	ePS	C	19 30	LmH B 19 8.5/um 6.3
	ePKKP	A	21 46	LmV B 18 9.6/um 6.4
	LmH	B	58.0	
	LmV	B	58.1	

Day	Phase		h m s	Remarks
16.	iPg	A	01 41 28.2	Vogtland, German Democratic Republic
	iSg	A	41 38.7	(MOX)
16.	eP	A	02 26 25.5	Molucca Passage 2.01 N 126.55 E
	ePP	A	30 43	H = 02 12 23.3 h = 47 km MB = 5.5
				D = 103.97 Az = 323.5 (NEIC)
16.	eP	A	22 02 20	Near East Coast of Honshu, Japan
	ePP	A	05 20	37.04 N 141.57 E
	e	A	05 27	H = 21 50 01.1 h = 40 km MB = 5.1
	LmH	B	37.1	D = 82.22 Az = 330.5 (NEIC)
	LmV	B	46.1	PV A 1.5s 27.6nm M = 5.1
				LmH B 16.5 0.7/um 5.1
				LmV B 15 0.8/um 5.2
17.	eP	A	00 22 16	Kurile Islands 46.17 N 153.27 E
				H = 00 10 19.0 h = normal MB = 4.3
				D = 78.02 Az = 336.2 (NEIC)
17.	+iPKP	AB	05 16 24.7	New Hebrides Islands 19.38 S 169.36 E
	eipPKP	AB	17 13	H = 04 57 12.7 h = 194 km MB = 6.0
	-iPP	AB	19 47.7	D = 144.15 Az = 335.7 (NEIC)
	e	A	27 53	PKIKPV A 1.2s 311.0nm
				PKIKPV B 6 1.42/um
				PPV A 1.5 186.0nm M = 6.0
17.	eP1	AB	08 44 02.5	Luzon, Philippine Islands
	eiP2	A	44 17	13.37 N 122.79 E
	eiP3	AB	44 25	H = 08 30 51.8 h = normal MB=5.6 MS=7.0
	eSKS3	B	54 35	D = 92.67 Az = 323.5 (NEIC)
	eS3	B	55 30	P1V A 2.0s 59.8nm M = 5.7
	e	B	56 16	P2V A 1.7 218.0nm 6.3
	e	B	57 33	P3V A 2.7 930.0nm 6.7
	LmH	B	09 26.6	P3V B 4 2.4/um 7.0
	LmV	B	26.8	LmH B 17 265.0/um 7.8
				LmV B 18 208.5/um 7.7

Day	Phase		h m s	Remarks
17.	e(Sg)	A	08 59 07	Czechoslovakia 50.18 N 14.40 E
				Explosion 11 t
				H = 08 58 01 (PRU)
				D = 1.84 Az = 286 (ISC)
17.	eP	A	12 27 09.5	Tadzhikistan-Sinkiang Border Region
				38.35 N 74.08 E
				H = 12 19 05.9 h = 134 km MB = 4.6
				D = 44.93 Az = 307 (ISC)
17.	ePKP	A	13 56 49	Tonga 16.21 S 174.92 W
				H = 13 37 10.2 h = 0 km MB = 4.8
				D = 145.25 Az = 353 (ISC)
17.	eP	A	15 54 06	Southern Sumatra 5.19 S 103.24 E
	epP	A	54 27	H = 15 40 48.0 h = 68 km MB = 5.7
	ePP	A	57 57	D = 95.01 Az = 320.3 (NEIC)
	eSKS	B	16 04 30	h = 81 km
	eS	B	05 12	PV A 1.2s 20.3nm M = 5.4
	LmH	B	38.6	LmH B 21 0.8/um
	LmV	B	45.9	LmV B 18 0.6/um
17.	eP	A	16 52 12	Southwest of Sumatra 6.83 S 102.07 E
	e	A	52 39	H = 16 38 49.3 h = 60 km MB = 5.2
				D = 95.52 Az = 320.2 (NEIC)
18.	eSg	A	02 06 49.5	Poland 50.33 N 18.82 E
				H = 02 04 27.7 M = 2.6 (WAR)
				D = 4.61 Az = 277 (ISC)
18.	ePKP	A	02 51 08	Tonga Islands 15.08 S 175.18 W
				H = 02 31 32.5 h = normal MB = 4.9
				D = 144.11 Az = 352.6 (NEIC)
18.	eP	AB	11 20 16.5	Molucca Passage 2.03 N 126.58 E
	ePm	A	20 25	H = 11 06 14.7 h = normal MB=6.0 MS=6.5
	e	AB	20 38	D = 103.97 Az = 323.5 (NEIC)
	ePP	B	24 40	PV A 1.3s 21.8nm M = 5.8
	eX	B	25 00	PVm A 1.8 71.0nm 6.2

Day	Phase		h m s	Remarks
cont.				
18.	iSKS	B	11 30 53	PV B 7s 0.76/um M = 6.6
	iS	B	32 00	PPV B 9 1.2/um 6.4
	eSP	B	33 28	XV B 14 1.9/um 6.5
	ePKKP	A	36 10	LmH B 18.5 17.5/um 6.6
	eSS	C	39 15	LmV B 17 13.0/um 6.5
	LmH	B	12 05.4	
	LmV	B	12.4	
18.	ePKIKP	A	20 04 52	<u>South of Fiji Islands</u> 22.14 S 179.65 W
	ePKHKP	A	04 56.5	H = 19 46 08.9 h = 566 km MB = 4.9
	ePKP2	A	05 06	D = 150.21 Az = 345.5 (NEIC)
				PKPV onsets traces
18.	eX	A	21 38 59	XV A 1.5s 17.6nm
19.	-eiP	AB	11 52 46.5	<u>Near Islands, Aleutian Is.</u>
	epP	AB	53 07	52.84 N 173.77 E
	e	A	54 15	H = 11 41 07.8 h = 81 km MB = 5.8
	e	A	55 25	D = 75.80 Az = 348.4 (NEIC)
	ePP	B	55 40	h = 80 km
	epPP	B	56 00	PV A 1.4s 151.0nm M = 5.7
	eS	B	12 02 20	LmH B 18 0.7/um
	eSP	B	02 48	LmV B 21 0.7/um
	ePS	B	03 08	
	LmV	B	25.0	
	LmH	B	28.2	
19.	ePP	A	12 25 33	<u>Turkey</u> 40.0 N 40.3 E
				H = 12 20 07.9 h = 33 km MB = 4 (ISC)
				D = 22.65
19.	eP	A	17 15 25	<u>Near East Coast of Honshu, Japan</u>
				36.67 N 140.97 E
				H = 17 03 08.4 h = 60 km MB = 4.6
				D = 82.31 Az = 330.2 (NEIC)
				traces

Day	Phase		h m s	Remarks
20.	ePKHKP	A	01 55 52.5	<u>Tonga</u> 22.0 S 175.22 W
				H = 01 35 58.0 h = 0 km MB = 5.1
				D = 150.90 Az = 351 (ISC)
20.	eP	A	07 11 25.5	<u>Fox Islands, Aleutian Is.</u>
				52.92 N 166.74 W
				H = 06 59 33.2 h = 16 km MB = 4.7
				D = 76.81 Az = 1.1 (NEIC)
20.	e	A	09 25 40	<u>Central Mid-Atlantic Ridge</u>
				8.92 N 39.68 W
				H = 09 15 31.6 h = normal MB = 4.7
				D = 59.19 Az = 35.3 (NEIC)
20.	eP dif	A	19 23 02	<u>Sumbawa Island Region</u> 8.27 S 117.38 E
	ePKIKP	A	27 16	H = 19 09 06.6 h = 162 km MB = 5.7
	ePP	B	28 00	D = 106.34 Az = 320.3 (NEIC)
	eiSKS	B	33 26	PKIKPV A 1.3s 19.6nm
	ePS	B	35 55	PKKPV A 1.4 18.6nm
	ePKKP	A	38 35.5	LmH B 21.5 2.8 μm
	e	A	38 53	LmV B 16 1.2 μm
	LmH	B	20 06.8	
	LmV	B	23.4	
21.	eP	A	07 27 29	<u>Eastern Mediterranean Sea</u>
				33.95 N 32.36 E
				H = 07 22 25.2 h = 0 km
				D = 22.54 Az = 324 (ISC)
21.	ePKP	A	09 32 59.5	<u>Tonga Islands</u> 16.13 S 173.70 W
				H = 09 13 29.7 h = 86 km MB = 4.6
				D = 145.31 Az = 354.0 (NEIC)
21.	e	A	12 34 12	<u>Crete</u> 35.0 N 23.5 E
				H = 12 29 50 (BCIS)
				D = 17.87

Day	Phase		h m s	Remarks
21.	eP	A	21 57 48.5	Kurile Islands 48.57 N 153.67 E
				H = 21 46 12.2 h = 105 km MB = 4.9
				D = 75.94 Az = 336.2 (NEIC)
				PV A 1.4s 30.2nm M = 4.9
22.	eP	A	01 17 02	Tibet 28.14 N 86.99 E
	e	A	17 17.5	H = 01 06 57.2 h = normal MB = 5.2
	LmH	B	40.4	D = 59.76 Az = 314.5 (NEIC)
	LmV	B	45.6	PV A 1.9s 37.8nm M = 5.2
				LmH B 26 1.7/um 5.1
				LmV B 13 0.7/um 5.0
22.	eP	A	06 21 54	Burma-China Border Region
	e	A	21 58.5	22.02 N 100.62 E
	LmH	B	54.0	H = 06 10 28.3 h = normal MB=4.7 MS=5.4
	LmV	B	56.7	D = 72.68 Az = 318.2 (NEIC)
				LmH B 20s 1.0/um M = 5.1
				LmV B 13 0.3/um 4.8
22.	eP	A	14 11 24	Leeward Islands 15.31 N 61.31 W
				H = 14 00 43.5 h = 156 km MB = 5.1
				D = 67.50 Az = 41.2 (NEIC)
				PV A 0.8s 13.5nm M = 4.8
22.	eP	A	21 10 32.5	Andreanof Islands, Aleutian Is.
				51.18 N 179.24 W
				H = 20 58 36.0 h = 40 km MB = 4.9
				D = 78.13 Az = 353.0 (NEIC)
				PV A 1.3s 19.7nm M = 5.0
22.	eP	A	22 52 54.5	Mindanao, Philippine Islands
				5.95 N 126.32 E
				H = 22 39 15.5 h = 102 km MB = 5.1
				D = 100.68 Az = 323.9 (NEIC)
23.	eSg	A	02 04 09.5	Poland 50.37 N 18.85 E
				H = 02 01 38.8 M = 3.0 (WAR)
				D = 4.6

Day	Phase		h m s	Remarks
23.	ePP	A	02 12 51	Southern Greece 36.2 N 22.0 E
				H = 02 08 39 (BCIS)
				D = 16.2
				traces
23.	e	A	04 09 04	Central Italy 43.4 N 13.1 E
	ePg	A	09 22	H = 04 07 02 (BCIS)
	eSn	A	10 12	D = 7.35
23.	ePKHKP	A	06 10 41.5	Loyalty Islands Region 22.76 S 170.25 E
	e	A	10 56	H = 05 50 59.8 h = normal MB = 5.5
				D = 147.55 Az = 334.4 (NEIC)
23.	eP1	AB	07 07 26	Near Islands, Aleutian Is.
	eP2	AB	07 30	51.30 N 174.22 E
	ePP2	A	10 27	H = 06 55 33.2 h = 27 km MB=5.8 MS=5.9
	eS	B	17 20	D = 77.36 Az = 348.7 (NEIC)
	LmH	B	44.7	P1V A 1.2s 54.9nm M = 5.5
	LmV	B	51.0	P2V A 2.0 188.0nm 5.8
				P2V B 10 1.3/um 5.9
				LmH B 18.5 3.7/um 5.7
				LmV B 17.5 3.5/um 5.8
23.	eP	A	15 25 40	Northwest Territories, Canada
				66.74 N 130.00 W
				H = 15 15 38.6 h = normal MB = 4.8
				D = 59.37 Az = 27.3 (NEIC)
23.	eP	A	19 25 33	Szechwan Province, China
				31.88 N 100.06 E
				H = 19 14 53.1 h = normal MB=5.3 MS=5.4
				D = 65.17 Az = 315.5 (NEIC)
				PV A 1.6s 24.8nm M = 5.3
23.	+iP	AB	19 55 06.5	Ryukyu Islands 29.27 N 130.41 E
	eS	B	20 05 46	H = 19 42 38.8 h = 34 km MB=5.9 MS=5.5
	LmH	B	37.4	D = 83.85 Az = 325.9 (NEIC)

Day	Phase		h m s	Remarks
cont. 23.	LmV	B	20 37.5	PV A 1.5s 264.0nm M = 6.1 LmH B 15 14.5 /um 6.5 LmV B 15.5 14.5 /um 6.5
23.	eP	A	21 53 41.5	<u>Talaud Islands</u> 4.19 N 126.83 E
	epP	A	53 51	H = 21 39 47.6 h = 35 km MB = 5.6
	ePP	A	57 59	D = 102.39 Az = 323.9 (NEIC) h = 33 km pPV A 1.5s 30.1nm
24.	eP	A	00 38 52	<u>Burma</u> 21.62 N 98.32 E H = 00 27 32.7 h = normal MB = 4.9 D = 71.55 Az = 317.9 (NEIC)
24.	eP	A	00 47 19.5	<u>South of Honshu, Japan</u>
	e	A	47 30	31.72 N 139.30 E
	e	A	47 40	H = 00 34 36.8 h = 20 km MB=5.6 MS=5.6
	ePP	A	50 38	D = 85.89 Az = 329.7 (NEIC)
	eS	C	57 50	PV A 1.4s 32.6nm M = 5.3
	LmH	B	01 29.3	LmH B 13.5 6.3 /um 6.2
	LmV	B	32.0	LmV B 16 3.3 /um 5.9
24.	eP	A	07 26 06.5	<u>Off East Coast of Kamchatka</u> 51.56 N 161.63 E H = 07 14 26.2 h = normal MB = 4.8 D = 75.04 Az = 340.8 (NEIC) PV A 1.8s 33.8nm M = 5.0
24.	ePn	A	11 27 45	<u>Austria</u> 47.3 N 11.4 E
	ePg	A	28 03.5	H = 11 27 02 (BCIS)
	eSn	A	28 30.5	D = 3.33
	eSg	A	28 47.5	
24.	ePn	A	11 46 11.5	<u>Austria</u> 47.3 N 11.4 E
	ePg	A	46 25.5	H = 11 45 21 (BCIS)
	eSn	A	46 50.5	D = 3.33
	eSg	A	47 06	

Day	Phase		h m s	Remarks
24.	eSg	A	15 56 10	Poland 50.32 N 18.88 E
				H = 15 53 39.5 M = 2.4 (WAR)
				D = 4.65
25.	eP	A	03 14 06.5	North Atlantic Ocean 57.49 N 33.24 W
				H = 03 08 26.5 h = normal MB = 4.2
				D = 26.78 Az = 85.3 (NEIC)
				traces
25.	+iP	A	09 07 55	Kurile Islands 50.24 N 156.86 E
	LmV	B	45.7	H = 08 56 15.2 h = 40 km MB = 5.4
	LmH	B	45.8	D = 75.21 Az = 338.0 (NEIC)
				PV A 1.7s 91.0nm M = 5.5
				LmH B 16.5 2.0 μm 5.5
				LmV B 17.5 1.0 μm 5.2
25.	eP	A	13 46 47.5	Kurile Islands 45.97 N 151.12 E
				H = 13 34 58.6 h = 78 km MB = 4.7
				D = 77.59 Az = 335.0 (NEIC)
25.	eP1	A	22 54 50.5	Gulf of California 25.89 N 109.88 W
	eP2	A	54 55	H = 22 42 02.8 h = normal MB=5.5 MS=5.5
	eS	B	23 05 40	D = 87.98 Az = 32.9 (NEIC)
	ePS	B	06 36	P1V A 1.0s 19.7nm M = 5.4
	LmV	B	34.5	P2V A 1.1 36.3nm 5.6
	LmH	B	34.6	LmH B 18 4.5 μm 5.9
				LmV B 19.5 6.1 μm 6.1
26.	eP	A	02 41 52.5	Philippine Sea 21.64 N 126.33 E
				H = 02 29 05.1 h = normal MB = 4.8
				D = 87.99 Az = 324.6 (NEIC)
				traces
26.	eP1	A	02 50 00	Southwestern Ryukyu Islands
	eP2	A	50 04.5	23.42 N 123.96 E
	eS	C	03 00 28	H = 02 37 21.3 h = 16 km MB=5.5 MS=6.0
	LmH	B	25.8	D = 85.27 Az = 323.8 (NEIC)
	LmV	B	33.6	P1V A 1.6s 33.0nm M = 5.3

Day	Phase		h m s	Remarks
cont. 26.				P2V A 1.6s 63.2nm M = 5.6
				LmH B 16.5 12.9/um 6.4
				LmV B 16 9.3/um 6.3
26.	ePKHKP	A	05 28 31.5	Fiji Islands Region 21.28 S 179.08 W
	ePKP2	A	28 38.5	H = 05 09 50.5 h = 634 km MB = 4.9
				D = 149.51 Az = 346.5 (NEIC)
				PKHKPV A 1.2s 16.3nm
				PKP2V A 1.4 18.6nm
26.	eP	A	08 06 46.5	Tadzhik-Sinkiang Border Region
	epP	A	07 15.5	38.33 N 73.93 E
	esP	A	07 26.5	H = 07 58 42.7 h = 123 km MB = 5.3
	LmH	B	18.3	D = 44.85 Az = 306.9 (NEIC)
				h = 126 km
26.	eP	A	08 41 35	Tadzhik-Sinkiang Border Region
				38.29 N 73.90 E
				H = 08 33 30.6 h = 120 km MB = 5.2
				D = 44.86 Az = 307.0 (NEIC)
26.	ePKIKP	A	11 22 30	La Rioja Province, Argentina
				28.05 S 68.94 W
				H = 11 04 19.0 h = 117 km MB = 5.7
				D = 105.57 Az = 40.7 (NEIC)
26.	LmH	B	11 32.8	
26.	eP	A	15 08 33.5	Off East Coast of Honshu, Japan
				36.64 N 141.98 E
				H = 14 56 13.0 h = 52 km MB = 5.1
				D = 82.73 Az = 330.7 (NEIC)
26.	eP	A	17 47 38	Off East Coast of Honshu, Japan
	LmV	C	18 31.7	36.62 N 142.02 E
				H = 17 35 17.3 h = 49 km MB = 5.1
				D = 82.76 Az = 330.7 (NEIC)
				PV A 1.3s 17.5nm M = 5.1
				LmV C 15 0.5/um 5.0

Day	Phase		h m s	Remarks
26.	eP	A	21 59 31	Near Islands, Aleutian Is. 52.82 N 173.82 E H = 21 47 53.7 h = 102 km MB = 5.0 D = 75.82 Az = 348.4 (NEIC)
27.	LmV LmH	C C	00 20.5 20.7	Probably Azores Islands Region 36.31 N 33.91 W H = 00 01 05.0 h = normal MB = 4.3 (NEIC) D = 35.5 LmV C 18s 0.35/um M = 4.3
27.	e e(PP)	A A	02 21 06 23 53	Near S. Coast of Honshu, Japan 35.48 N 139.92 E H = 02 08 31.4 h = 64 km MB = 5.0 D = 82.90 Az = 329.8 (NEIC)
27.	eP	A	03 15 54.5	Crete 34.4 N 26.4 E H = 03 11 28 h = 30 km MB = 4.0 (ISC) D = 19.49
27.	eP	A	03 50 08.5	Southwestern Ryukyu Islands 24.32 N 123.56 E H = 03 37 42.8 h = 77 km MB = 5.0 D = 84.33 Az = 323.6 (NEIC)
27.	eP ePS LmH LmV	AB B B B	12 43 49 54 20 13 15.0 15.0	Near Islands, Aleutian Is. 52.58 N 172.87 E H = 12 32 05.1 h = 43 km MB=5.6 MS=5.2 D = 75.94 Az = 347.8 (NEIC) PV A 1.5s 110.5nm M = 5.6 LmH B 22 1.3/um 5.2 LmV B 22 2.9/um 5.6
27.	eP	A	15 41 42.5	Greece 38.43 N 21.09 E H = 15 38 24.2 h = normal MB = 4.1 D = 13.95 Az = 334.2 (NEIC)

Day	Phase		h m s	Remarks
28.	eP	A	03 23 38	Jan Mayen Islands Region 72.3 N 1.2 W
				H = 03 18 35.7 h = 0 km
				D = 22.46 Az = 158 (ISC)
				PV A 1.3s 17.5nm
28.	eP	A	03 42 43.5	Hokkaido, Japan Region 42.57 N 144.03 E
	e	A	43 12	H = 03 30 48.9 h = 65 km MB = 4.5
				D = 78.31 Az = 331.3 (NEIC)
28.	eP	A	03 43 53.5	Southern Iran 28.57 N 52.67 E
	eS	B	49 48	H = 03 36 38.2 h = 37 km MB=5.2 MS=4.4
	LmH	B	04 05.0	D = 37.89 Az = 317.1 (NEIC)
	LmV	B	05.0	PV A 1.0s 15.7nm M = 4.8
				LmH B 15.5 0.7 μm 4.5
				LmV B 12 0.5 μm 4.7
28.	ePKP	A	07 42 58.5	Loyalty Islands Region 21.82 S 170.31 E
				H = 07 23 23.5 h = 58 km
				D = 146.73 Az = 335.1 (NEIC)
28.	eP	A	10 53 55.5	Ethiopia 11.82 N 42.70 E
	LmV	C	11 13.5	H = 10 45 30.3 h = normal MB = 4.6
	LmH	C	14.0	D = 46.31 Az = 333.0 (NEIC)
				PV A 1.4s 14.0nm M = 4.7
28.	eP	A	11 05 10	Ethiopia 14 N 42 E
				H = 10 57 05 MB = 4.0 (NORSAR)
				D = 44.0
28.	eP	A	13 43 30.5	Ethiopia 11.67 N 42.83 E
	e	A	45 05.5	H = 13 35 04.5 h = normal MB = 5.0
				D = 46.50 Az = 332.9 (NEIC)
				PV A 1.5s 37.7nm M = 5.1
28.	eP1	AB	13 50 32.5	Ethiopia 11.74 N 42.75 E
	e	A	50 35.5	H = 13 42 06.7 h = normal MB=5.3 MS=5.3
	ePP	C	52 24	D = 46.40 Az = 333.0 (NEIC)
	eS	B	57 22	P1 A 1.4s 48.8nm M = 5.3

Day	Phase		h m s	Remarks
cont. 28.	LQ	C	14 00.5	SH B 16s 2.3/um M = 5.8
	LmV	B	13.8	LmH B 14.5 3.4/um 5.4
	LmH	B	15.5	LmV B 12.5 2.1/um 5.4
28.	eP	A	14 08 42	Ethiopia 11.5 N 42.64 E
				H = 14 00 32 h = 187 km MB = 4.0
				D = 46.56 Az = 333 (ISC)
28.	eP	A	14 27 18.5	Ethiopia 11.70 N 42.93 E
	e	A	27 22	H = 14 18 52.3 h = normal MB=5.4 MS=5.4
				D = 46.51 Az = 332.9 (NEIC)
				PV A 1.4s 74.5nm M = 5.5
28.	e	A	14 45 17	
28.	eX	A	14 57 34	XV 1.3s 19.7nm
28.	eP	A	15 07 32	Ethiopia 11.75 N 42.87 E
				H = 14 59 06.7 h = normal MB = 5.2
				D = 46.45 Az = 332.9 (NEIC)
				PV A 2.0s 167.0nm M = 5.7
28.	ePKIKP	A	17 04 05	New Guinea 4.53 S 141.82 E
	epPKIKP	A	04 28	H = 16 45 25.3 h = 81 km MB = 5.5
				D = 118.04 Az = 326.6 (NEIC)
				h = 85 km
28.	eSg	A	20 32 25	Poland 50.34 N 18.83 E
				H = 20 29 57.9 M = 3.0 (WAR)
				D = 4.6
28.	ePKP2	A	22 46 36	Fiji Islands Region 21.87 S 179.53 W
				H = 22 27 43.1 h = 610 km MB = 4.8
				D = 149.97 Az = 345.8 (NEIC)
28.	eP	AB	24 08 22	Southwestern Ryukyu Islands
	eS	C	18 50	23.34 N 123.76 E
	eSS	C	25 06	H = 23 55 47.3 h = normal MB=5.3 MS=6.0

Day	Phase		h m s	Remarks
cont. 28.	LmH	B	24 45.5	D = 85.23 Az = 323.7 (NEIC)
	LmV	B	53.1	PV A 1.7s 60.6nm M = 5.5
				LmH B 15.5 6.9 /um 6.2
				LmV B 15 5.4 /um 6.1
29.	e	A	06 33 44.5	Southern Sumatra 2.15 S 99.72 E
				H = 06 20 37.6 h = 50 km MB = 5.2
				D = 90.44 Az = 320.5 (NEIC)
29.	ePKIKP	A	11 42 41	South of Fiji Islands 23.55 S 179.73 W
	ePKHKP	A	42 58	H = 11 24 03.9 h = 535 km MB = 4.7
	e	A	43 32.5	D = 151.55 Az = 344.8 (NEIC)
				PKIKPV traces
				PKHKPV A 1.4s 14.0nm
29.	eX	A	23 53 36.5	XV A 1.4s 18.6nm
30.	eP1	A	00 12 22.5	Svalbard Region 76.25 N 5.99 E
	eP2	AB	12 28	H = 00 06 52.5 h = normal MB=5.1 MS=5.4
	eS	B	17 00	D = 25.80 Az = 171.8 (NEIC)
	LmH	B	22.8	P1V A 1.6s 33.0nm M = 4.7
	LmV	B	22 22.8	P2V A 1.5 196.0nm 5.5
				P2H A 1.6 175.0nm 5.5
				LmH B 18 1.8 /um 4.7
				LmV B 16 2.1 /um 4.9
30.	eP	A	03 13 48.5	Off East Coast of Honshu, Japan
	LmH	B	47.8	40.76 N 143.07 E
	LmV	B	52.5	H = 03 01 45.5 h = 54 km MB = 4.8
				D = 79.55 Az = 331.0 (NEIC)
				PV A 2.0s 59.9nm M = 5.2
				LmH B 17 0.9 /um
				LmV B 16 0.7 /um
31.	eP	A	20 57 11	Kurile Islands 50.18 N 157.05 E
				H = 20 45 31.3 h = 44 km MB = 4.8
				D = 75.31 Az = 338.1 (NEIC)

Day	Phase	h m s	Remarks
31.	eP A	23 36 45	Rumania 45.72 N 26.66 E H = 23 34 07.9 h = 159 km MB = 4.1 D = 11.18 Az = 301.6 (NEIC)

Day	Phase		h m s	Remarks
1.	eP	A	06 37 53.5	<u>Ethiopia</u> 11.61 N 42.93 E
	e	A	38 00	H = 06 29 27.7 h = normal MB = 4.8
	ePcP	A	39 24.5	D = 46.59 Az = 332.9 (NEIC)
1.	eP1	A	07 21 03	<u>Ethiopia</u> 11.66 N 43.04 E
	eP1m	AB	21 08	H = 07 12 37.0 h = 31 km MB=5.5 MS=5.9
	eP2	A	22 10.5	D = 46.60 Az = 332.8 (NEIC)
	eiS1	C	27 56	P1Vm A 1.8s 77.6nm M = 5.4
	eS2	C	29 00	P2V A 1.4 51.2nm 5.3
	LmV	B	46.5	S2H C 17 4.3 μm 5.9
	LmH	B	47.2	LmH B 15 4.2 μm 5.5
				LmV B 17 2.7 μm 5.4
1.	e	A	07 29 31	
1.	eP	A	07 47 07	<u>Ethiopia</u> 11.65 N 42.93 E
	e	A	47 12	H = 07 38 40.5 h = normal MB = 4.9
				D = 46.57 Az = 332.9 (NEIC)
				PV A 1.8s 40.6nm M = 5.1
1.	e	A	07 50 02.5	<u>Ethiopia</u> 11.82 N 42.79 E
	e	A	50 10.5	H = 07 41 21.46 h = normal MB = 4.7 (NEIC)
				D = 46.35
1.	eP	A	08 53 52.5	<u>Iceland Region</u> 68.17 N 18.90 W
				H = 08 48 51.7 h = normal MB = 4.3
				D = 23.03 Az = 124.3 (NEIC)
				traces
1.	eP	A	08 56 12	<u>Iceland Region</u> 67.76 N 18.56 W
	Pm	A	56 18	H = 08 51 13.0 h = normal MB = 4.5
	LmH	B	09 05.6	D = 22.69 Az = 123.9 (NEIC)
				PV A 1.4s 20.9nm M = 4.4
				PmV A 2.0 77.0nm 4.8
2.	eP	A	01 35 20	<u>Southern Iran</u> 27.09 N 61.56 E
	e	A	35 30	H = 01 27 09.0 h = normal MB = 5.0
				D = 44.39 Az = 315.8 (NEIC)

Day	Phase		h m s	Remarks
2.	eP	A	02 51 19	<u>Afghanistan-USSR Border Region</u>
	e	A	51 23	37.70 N 69.98 E
	e	A	51 50.5	H = 02 43 24.8 h = 54 km MB = 4.4
	LmH	B	03 09.7	D = 42.74 Az = 307.0 (NEIC)
	LmV	B	12.9	LmH B 16.5s 1.5 µm
				LmV B 14 0.8 µm
2.	eP	A	19 34 51	<u>Central Mid-Atlantic Ridge</u>
	LmH	B	55.3	7.25 N 34.40 W
	LmV	B	56.1	H = 19 25 00.4 h = normal MB=5.1 MS=5.0
				D = 57.63 Az = 32.8 (NEIC)
				PV A 2.0s 51.3nm M = 5.2
				LmH B 22 1.7 µm 5.1
				LmV B 22 2.1 µm 5.3
3.	-iP	A	11 02 21.6	<u>Kurile Islands</u> 45.85 N 151.12 E
				H = 10 50 26.3 h = normal MB = 5.0
				D = 77.69 Az = 335.0 (NEIC)
				PV A 1.0s 19.7nm M = 5.1
3.	-iP	AB	14 06 19.8	<u>Colombia</u> 4.69 N 75.63 W
	epP	AB	06 57.5	H = 13 54 01.8 h = 158 km MB = 6.2
	esP	AB	07 12	D = 84.65 Az = 39.7 (NEIC)
	ePP	B	09 32	h = 154 km
	esPP	B	10 26	PV A 2.0s 325.0nm M = 5.8
	eiS	B	16 31	SH B 15.5 4.2 µm 6.2
	e	B	17 24	LmH B 20 1.6 µm
	eSS	B	22 12	LmV B 20 1.4 µm
	LmV	B	39.3	
	LmH	B	39.5	
3.	eP	A	17 35 58	<u>Southern Sumatra</u> 4.62 S 103.19 E
				H = 17 22 46.3 h = 100 km MB = 5.2
				D = 94.54 Az = 320.3 (NEIC)
				traces

Day	Phase		h m s	Remarks
3.	eP	A	20 14 45	<u>Central Mid-Atlantic Ridge</u> 7.46 N 36.26 W H = 20 04 47.7 h = normal MB=5.1 MS=4.7 D = 58.47 Az = 33.6 (NEIC) PV A 1.9s 56.9nm M = 5.3 LmH C 22 0.7/um 4.7
	LmH	C	35.5	
	LmV	B	40.0	
4.	e(P)	A	01 42 15	<u>Ascension Island Region</u> 6.80 S 12.01 W H = 01 31 57.8 h = normal MB = 4.7 D = 60.75 Az = 17.0 (NEIC)
4.	eP	A	22 00 33	<u>Western Gulf of Aden</u> 12.16 N 46.19 E H = 21 51 58.4 h = normal MB = 4.5 D = 47.62 Az = 330.7 (NEIC) PV A 1.4s 23.2nm M = 5.0
4.	eP	A	22 02 55.5	<u>Kurile Islands</u> 43.37 N 147.73 E H = 21 50 53.9 h = normal MB = 5.2 D = 78.87 Az = 333.3 (NEIC) PV A 1.3s 28.4nm M = 5.1 LmH B 20 1.7/um 5.4 LmV B 15 1.5/um 5.5
	LmH	B	35.4	
	LmV	B	42.5	
4.	eP	A	24 08 45.5	<u>Kurile Islands</u> 43.46 N 147.78 E H = 23 56 43.5 h = normal MB = 5.3 D = 78.99 Az = 333.3 (NEIC) PV A 1.3s 26.2nm M = 5.1
5.	eP	A	02 07 47	<u>Western Gulf of Aden</u> 12.25 N 46.42 E H = 01 59 12.6 h = normal MB = 4.6 D = 47.65 Az = 330.5 (NEIC) PV A 2.1s 47.9nm M = 5.1
	LmH	C	27.5	
5.	eP	A	04 45 30	<u>Panay, Philippine Islands</u> 11.94 N 122.50 E H = 04 32 08.1 h = 6 km MB=5.4 MS=5.4 D = 93.65 Az = 323.4 (NEIC) LmH B 17s 2.9/um M = 5.8 LmV B 18 2.8/um 5.8
	e	A	45 37	
	LmH	B	05 28.1	
	LmV	B	36.7	

Day	Phase		h m s	Remarks
5.	+eiP	A	22 29 00.7	<u>Kurile Islands</u> 43.61 N 147.71 E
	eS	C	39 00	H = 22 16 59.6 h = normal MB=5.4 MS=5.6
	eSS	C	44.0	D = 78.65 Az = 333.2 (NEIC)
	LmH	B	23 01.4	PV A 1.2s 50.8nm M = 5.4
	LmV	B	08.7	LmH B 21 9.6/um 6.1
				LmV B 16 7.6/um 6.2
5.	eP	A	22 47 28	<u>Kurile Islands</u> 43.66 N 147.68 E
				H = 22 35 28.7 h = normal MB = 4.7
				D = 78.60 Az = 333.2 (NEIC)
5.	eP	A	23 18 52.5	<u>Kurile Islands</u> 43.51 N 147.76 E
				H = 23 06 51.13 h = normal MB = 4.5 (NEIC)
				D = 78.75
5.	eP	A	23 45 59.5	<u>Kurile Islands</u> 43.23 N 147.81 E
				H = 23 33 57.5 h = normal MB = 4.6
				D = 79.03 Az = 333.3 (NEIC)
				PV A 1.7s 24.2nm M = 4.9
6.	eP	A	00 13 57	<u>Kurile Islands</u> 43.65 N 147.65 E
	e	A	14 09.5	H = 00 01 56.4 h = normal MB = 5.3
	LmH	B	46.5	D = 78.59 Az = 333.2 (NEIC)
	LmV	B	53.5	PV A 2.2s 120.0nm M = 5.5
				LmH B 18 1.6/um 5.4
6.	eP	A	01 58 13	<u>Andreanof Islands, Aleutian Is.</u>
				51.42 N 178.44 W
				H = 01 46 18.2 h = 50 km MB = 5.0
				D = 77.96 Az = 353.5 (NEIC)
				PV A 1.7s 24.2nm M = 5.0
6.	eP	A	02 00 00	<u>Kurile Islands</u> 43.67 N 147.76 E
	eS	C	10 00	H = 01 48 00.3 h = normal MB=5.4 MS=5.8
	LmH	B	32.7	D = 78.61 Az = 333.3 (NEIC)
	LmV	B	39.6	PV A 1.3s 59.0nm M = 5.4
				LmH B 20 4.6/um 5.8
				LmV B 16 5.1/um 6.0

Day	Phase		h m s	Remarks
6.	e	A	05 53 07	Kurile Islands 43.42 N 147.69 E H = 05 40 52.6 h = normal MB = 4.1 D = 78.81 Az = 333.2 (NEIC)
6.	eP	A	10 56 03	Kurile Islands 43.74 N 147.62 E H = 10 44 04.8 h = normal MB = 4.2 D = 78.51 Az = 333.2 (NEIC)
6.	eP LmH LmV	AB B B	14 18 17.5 27.0 27.0	Crete 34.43 N 25.25 E H = 14 13 54.2 h = 16 km MB = 5.1 D = 19.01 Az = 332.6 (NEIC) PV A 0.7s 57.5nm M = 4.9 LmH B 15 1.6/um 4.5 LmV B 13 1.6/um 4.7
6.	eP e	A A	14 57 51 58 04.5	Kurile Islands 43.83 N 147.64 E H = 14 45 53.1 h = normal MB = 4.3 D = 78.43 Az = 333.2 (NEIC) PV A traces
6.	eP LmH	A C	15 10 02 49.5	Kurile Islands 43.48 N 147.61 E H = 14 58 01.7 h = normal MB = 4.9 D = 78.73 Az = 333.2 (NEIC)
6.	eP	A	15 21 05	Kurile Islands 43.55 N 147.72 E H = 15 09 04.4 h = normal MB = 4.5 D = 78.70 Az = 333.2 (NEIC)
6.	e	A	16 29 20.5	
6.	eP	A	18 40 54	Andreanof Islands, Aleutian Is. 51.77 N 173.48 W H = 18 28 56.9 h = 21 km MB = 4.5 D = 77.87 Az = 356.7 (NEIC) PV A 1.1s 12.1nm M = 4.9

Day	Phase		h m s	Remarks
6.	eP	A	18 58 09	Andreanof Islands, Aleutian Is.
				51.59 N 173.35 W
				H = 18 46 13.7 h = 40 km MB = 4.7
				D = 78.05 Az = 356.8 (NEIC)
				PV A 1.2s 16.3nm M = 4.9
6.	ePKIKP	A	19 01 27	Fiji Islands Region 19.98 S 176.45 W
	ePKHKP	A	01 31	H = 18 42 10.8 h = 242 km MB = 4.8
	ePKP2	A	01 35	D = 148.74 Az = 350.1 (NEIC)
				PKIKPV A traces
				PKHKPV A 1.8s 54.0nm
7.	ePKP	A	02 31 57.5	Fiji Region 16.48 S 178.71 W
				H = 02 12 23.0 h = 0 km MB = 7.4
				D = 144.92 Az = 349 (ISC)
7.	-iP	AB	03 12 56.5	Nicobar Islands Region 6.97 N 91.38 E
	e	B	15 10	H = 03 00 58.8 h = normal MB=5.9 MS=6.6
	+eS	B	22 52	D = 78.16 Az = 320.2 (NEIC)
	+iSS	B	28 04	PV A 1.5s 251.0nm M = 6.0
	iSSS	B	31 33	PV B 16 1.8/um 5.8
	LmH	B	51.3	SH B 18 13.5/um 7.8
	LmV	B	55.4	LmH B 19.5 23.2/um 6.5
				LmV B 17 20.3/um 6.6
7.	eP	A	03 26 45	Nicobar Islands Region 6.87 N 91.38 E
				H = 03 14 46.9 h = normal MB = 4.8
				D = 78.23 Az = 320.2 (NEIC)
				PV A 1.3s 24.0nm M = 5.1
7.	ePP1	A	12 41 55	Southwestern Atlantic Ocean
	ePP2	AB	42 10	58.30 S 13.39 W
	ePS	C	51 30	H = 12 22 47.26 h = normal
	eSS	C	57 22	MB = 6.2 MS = 6.7 (NEIC)
	LmV	B	13 27.4	D = 110.50
	LmH	B	28.4	PP2V A 2.3s 116.0nm M = 6.2
				LmH B 17 4.4/um 6.1
				LmV B 16 4.2/um 6.1

Day	Phase		h m s	Remarks
7.	eP	A	17 45 09.5	Ethiopia 11.69 N 43.02 E
	eS	B	52 00	H = 17 36 42.8 h = normal MB = 4.7
	LmH	B	18 10.3	D = 46.57 Az = 332.8 (NEIC)
	LmV	B	10.6	PV A 2.0s 51.3nm M = 5.1
				LmH B 14 1.1/um 5.0
				LmV B 17 0.9/um 4.9
7.	eP	A	18 09 32	Bering Sea 54.23 N 174.8 W
				H = 17 57 44.9 h = 0 km MB = 3.9
				D = 75.36 Az = 356 (ISC)
7.	eP	A	19 32 45	Albania 41.50 N 19.98 E
	e	A	32 58.5	H = 19 30 06.9 h = 16 km MB = 4.7
	LmH	B	36.3	D = 10.83 Az = 330.5 (NEIC)
	LmV	B	37.6	PV A 1.0s 15.7nm M = 5.3
				LmH B 14.5 1.8/um 4.1
				LmV B 8 1.2/um
7.	ePP	A	23 08 37	Northern Chile 20.67 S 69.19 W
				H = 22 50 54.51 h = 95 km MB = 5.6 (NEIC)
				D = 100.09
				PPV A 2.0s 34.2nm M = 5.6
8.	eP	A	04 15 50.5	Greece-Albania Border Region
				39.97 N 20.61 E
				H = 04 12 53.7 h = 45 km MB = 3.8
				D = 12.40 Az = 332.4 (NEIC)
8.	e	A	07 06 27.5	
8.	ePKIKP1	AB	13 00 24.5	New Hebrides Islands 15.78 S 167.22 E
	ePKIKP2	AB	00 28	H = 12 41 02.0 h = 35 km MB=5.8 MS=6.4
	ePP	AB	03 32.5	D = 140.05 Az = 335.8 (NEIC)
	ePKS	AB	04 05	PKIKP2V A 1.8s 142.0nm
	ePS	B	13 44	PPV A 3.0 527.0nm M = 6.3
	eSS	B	21 50	PPV B 12 2.7/um 6.5
	eSSS	C	27 00	LmH B 21 9.1/um 6.5
	LmH	B	14 04.5	LmV B 21 8.5/um 6.5
	LmV	B	04.6	

Day	Phase		h m s	Remarks
8.	ePKIKP	A	21 08 22.5	South of Fiji Islands 23.88 S 177.04 W
	ePKHKP	A	08 29.5	H = 20 48 46.0 h = 122 km MB = 5.5
	eiPKP2	A	08 36	D = 152.44 Az = 348.0 (NEIC)
	epPKIKP	A	08 57	h = 124 km
	epPKHKP	A	09 03.5	LmV C 33s 0.3/um
	e	A	09 09.5	
	e	A	09 16	
	e	A	09 23.5	
	LmH	C	22 37.0	
	LmV	C	37.0	
8.	+iP	A	22 06 42	Kurile Islands 47.00 N 152.29 E
				H = 21 54 59.2 h = 104 km MB = 5.6
				D = 76.99 Az = 335.5 (NEIC)
				PV A 1.1s 282.0nm M = 6.0
8.	LmH	C	23 58.4	LmH C 22s 0.35/um
	LmV	C	59.0	LmV C 26 0.6/um
9.	eP	A	08 44 50	Northeast of Taiwan 26.82 N 125.30 E
	epP	A	45 41.5	H = 08 32 42.1 h = 186 km MB = 5.4
				D = 83.25 Az = 324.0 (NEIC)
				h = 216 km
				PV A 1.5s 37.7nm M = 4.9
				pPV A 2.0 85.5nm
9.	eP	A	08 46 40.5	North Atlantic Ridge 45.98 N 27.67 W
	eS	C	51 12	H = 08 41 02.1 h = normal MB = 4.8 (NEIC)
	LmV	B	56.5	D = 26.32
	LmH	B	56.6	PV A 2.0s 59.8nm M = 4.8
				LmH B 14 1.0/um 4.5
				LmV B 18 0.8/um 4.4
10.	ePKP	A	00 37 46	Fiji Region 19.7 S 177.2 W
	e	A	37 53	H = 00 18 06 h = 0 km MB = 4.9
				D = 148.34 Az = 349 (ISC)
				PKPV A 1.3s 15.3nm

Day	Phase		h m s	Remarks
10.	ePKIKP	A	01 28 30	Fiji Islands Region 20.40 S 176.32 W
	ePKHKP	A	28 35	H = 01 09 15.7 h = 265 km MB = 4.8
	ePKP2	A	28 39	D = 149.17 Az = 350.1 (NEIC)
	epPKHKP	A	29 32	h = 226 km
				PKIKPV A traces
				PKHKPV A 1.5s 22.6nm
10.	ePKIKP	A	07 54 29	Fiji Islands Region 20.66 S 178.44 W
	ePKHKP	A	54 34	H = 07 35 46.9 h = 550 km MB = 4.4
	ePKP2	A	54 40	D = 149.04 Az = 347.5 (NEIC)
				PKIKPV A traces
				PKHKPV A 1.5s 27.6nm
10.	eP	A	08 05 30	North Atlantic Ridge 20.89 N 45.76 W
				H = 07 56 10.4 h = normal MB = 4.2
				D = 53.54 Az = 41.8 (NEIC)
10.	eP	A	18 15 47.5	Afghanistan-USSR Border Region
				36.37 N 71.35 E
				H = 18 07 47.4 h = 125 km MB = 4.5
				D = 44.41 Az = 308.2 (NEIC)
10.	ePKP	A	20 14 32	Sumba Island Region 9.78 S 119.30 E
	ePP	A	14 48.5	H = 19 55 55.8 h = normal MB=5.8 MS=5.1
	e	A	15 45	D = 108.71 Az = 320.2 (NEIC)
	LmH	C	58.5	PPV A 1.7s 36.4nm M = 5.8
	LmV	B	21 04.0	LmH C 30 0.85/um 5.1
10.	ePKHKP	A	23 55 11	Tonga Islands 21.69 S 174.44 W
	ePKP2	A	55 20	H = 23 35 18.0 h = normal MB = 4.5
				D = 150.71 Az = 352.1 (NEIC)
11.	e	A	02 35 18.5	Oaxaca, Mexico 16.09 N 95.42 W
				H = 02 22 17.8 h = normal MB = 4.9
				D = 88.10 Az = 37.5 (NEIC)
11.	eP	A	04 38 57	Panay, Philippine Islands 11.82 N 122.31 E
				H = 04 25 40.6 h = 21 km MB = 5.0
				D = 93.63 Az = 323.4 (NEIC)

Day	Phase		h m s	Remarks
11.	e	A	05 04 40	Sweden 59.17 N 13.5 E
	e	A	05 51	H = 05 01 40 h = 102 km
	eSg	A	06 13	D = 8.61 AZ = 188 (ISC)
11.	e	A	09 41 43.5	Halmahera 0.83 S 127.54 E
	eSKS	C	52 16	H = 09 23 06.6 h = normal MB=5.6 MS=5.3
	e(PKKP)	A	57 21	D = 106.82 Az = 323.3 (NEIC)
	LmH	B	38.0	LmH B 17.5s 0.9/um M = 5.4
	LmV	B	38.3	LmV B 18 1.0/um 5.4
11.	LmV	B	13 49.5	Near Coast of Guerrero, Mexico
	LmH	B	53.1	16.91 N 100.26 W
				H = 12 51 13.5 h = normal MB = 4.9 (NEIC)
				D = 89.7
				LmH B 16s 0.3/um M = 4.8
				LmV B 15 0.35/um 4.9
11.	e	A	14 47 12	Southern Greece 37.20 N 21.30 E
	LmH	B	52.7	H = 14 43 27 h = 0 km
				D = 15.12 Az = 336 (ISC)
				LmH B 13s 0.6/um M = 3.9
12.	+iP	A	02 20 36	New Hebrides Islands 20.12 S 169.21 E
	+ipP	A	20 50	H = 02 01 05.3 h = 55 km MB = 4.8
				D = 144.77 Az = 335.1 (NEIC)
				h = 50 km
				PV A 1.2s 65.0nm
12.	eP	A	04 53 21.5	Ryukyu Islands 29.60 N 130.51 E
	LmH	B	05 29.4	H = 04 40 54.9 h = 37 km MB = 5.1
	LmV	B	34.7	D = 83.63 Az = 325.9 (NEIC)
				PV A 1.7s 24.2nm M = 5.0
				LmH B 17.5 0.9/um 5.2
				LmV B 17 0.9/um 5.2
12.	eP	A	05 15 13.5	Hokkaido, Japan Region 41.59 N 142.01 E
	epP	A	15 30	H = 05 03 18.8 h = 73 km MB = 5.4
				D = 78.43 Az = 330.3 (NEIC)
				h = 64 km
				PV A 1.1s 26.2nm M = 5.1

Day	Phase		h m s	Remarks
12.	eP	A	08 38 59	Hindu Kush Region 36.41 N 70.84 E
	e	A	39 48	H = 08 31 07.8 h = 189 km MB = 4.9
				D = 44.07 Az = 308.1 (NEIC)
12.	eP	A	12 15 43.5	Nicobar Islands Region 7.57 N 95.05 E
				H = 12 03 58.9 h = 229 km MB = 4.7
				D = 80.04 Az = 320.0 (NEIC)
12.	eP	A	12 28 54.5	Kurile Islands 49.98 N 155.99 E
				H = 12 17 15.0 h = 55 km MB = 5.2
				D = 75.24 Az = 337.4 (NEIC)
12.	+iP	AB	14 00 51.5	Kurile Islands 50.89 N 157.44 E
	ePP	B	03 39	H = 13 49 15.8 h = 52 km MB = 6.1
	eS	B	10 22	D = 74.74 Az = 338.2 (NEIC)
	eSKS	B	10 48	PV A 1.6s 368.0nm M = 6.1
	e	B	11 25	PV B 8 3.26/um 6.3
	LmH	B	37.6	PH A 1.9 309.0nm 6.2
	LmV	B	37.6	PPV B traces
12.	e	A	14 09 55	Southern California 34.28 N 116.32 W
				H = 13 57 02.8 h = 8 km MB = 4.4
				D = 83.78 Az = 30.3 (NEIC)
12.	ePKHKP	A	15 20 08.5	South of Fiji Islands 26 S 180 W
				H = 15 00 08 (NORSAR)
				D = 153.8
13.	eP	A	04 17 47	Mindanao, Philippine Islands
	e	A	17 53.5	5.47 N 126.37 E
				H = 04 03 56.9 h = 30 km MB = 5.3
				D = 101.09 Az = 323.9 (NEIC)
13.	ePn	A	08 15 08.5	Southern Italy 39.09 N 16.96 E
	eSn	A	17 42	H = 08 12 15.4 h = 45 km MB = 4.7
	LmH	B	20.6	D = 12.16 Az = 343.6 (NEIC)
	LmV	B	20.6	PV A 1.2s 16.3nm M = 5
				LmH B 11 1.7/um 5.3
				LmV B 13 1.6/um

Day	Phase		h m s	Remarks
13.	eP	A	14 22 26	Ethiopia 11.94 N 43.79 E
				H = 14 13 56.9 h = normal MB = 4.8
				D = 46.69 Az = 332.2 (NEIC)
				PV A 1.2s 24.4nm M = 5.1
13.	eP	A	19 27 49	Near East Coast of Kamchatka
	epP	A	28 03.5	53.97 N 161.74 E
				H = 19 16 24.4 h = 50 km MB = 5.1
				D = 72.79 Az = 340.6 (NEIC)
				h = 48 km
				PV A 1.2s 16.3nm M = 4.8
13.	ePKIKP	A	20 12 44	South of Fiji Islands 19.88 S 179.77 E
	eiPKHKP	A	12 48	H = 19 54 14.8 h = 665 km MB = 5.0
	e(pPKP)	A	15 20	D = 147.90 Az = 345.8 (NEIC)
				PKHKPV A 1.7s 63.6nm
13.	ePKIKP	A	22 55 39	Near Coast of Southern Chile
				52.46 S 72.04 W
				H = 22 36 39.3 h = 11 km MB = 5.1
				D = 124.53 Az = 50.2 (NEIC)
				PKIKPV A 1.0s 11.8nm
14.	ePKHKP	A	02 10 46	Tonga 18.50 S 175.01 W
				H = 01 51 09.9 h = 83 km MB = 4.5
				D = 147.49 Az = 352 (ISC)
14.	eP	A	03 14 48	Crete 34.49 N 24.25 E
	LmH	B	23.4	H = 03 10 30.6 h = 37 km MB = 4.3
	LmV	B	23.4	D = 18.59 Az = 334.1 (NEIC)
14.	eP	AB	08 46 41	Costa Rica 10.68 N 84.76 W
	e	A	46 50.5	H = 08 34 00.1 h = normal MB=5.7 MS=6.5
	ePP	C	50 00	D = 85.84 Az = 39.4 (NEIC)
	eSKS	B	57 00	PV A 2.0s 77.0 nm M = 5.6
	ePS	B	58 15	PV B 6 0.8 μm 6.1
	eSS	B	09 02 48	LmH B 22.5 33.1 μm 6.7
	eSSS	C	06 40	LmV B 18.5 19.2 μm 6.6

Day	Phase		h m s	Remarks
cont. 14.	eiLQ	C	09 09 15	
	eiLR	C	14 22	
	LmH	B	13.8	
	LmV	B	27.0	
14.	e	A	09 07 14	Costa Rica 10.71 N 84.76 W
				H = 08 54 29.2 h = normal MB = 4.8
				D = 85.82 Az = 39.4 (NEIC)
14.	ePKP2	A	15 27 37	Kermadec Islands 30.36 S 178.14 W
				H = 15 07 35.1 h = 282 km MB = 4.4 (NEIC)
				D = 158.45
14.	eP	A	16 31(18)	Mindanao, Philippine Islands
	LmH	B	17 20.2	6.64 N 124.09 E
	LmV	B	26.6	H = 16 17 42.1 h = 23 km MB=5.4 MS=4.9
				D = 98.81 Az = 323.5 (NEIC)
				PV A traces
				LmH B 18.5s 0.7/um M = 5.2
				LmV B 16 0.5/um 5.1
14.	eP	A	19 58 05	Kirgiz-Sinkiang Border Region
				41.6 N 77.8 E
				H = 19 49 44 M = 4.8
				D = 45.3 (ANUSSR)
				PV A 1.3s 13.1nm M = 4.6
14.	e	A	21 32 41	
15.	LmH	C	04 35.8	East New Guinea 6.03 S 146.08 E
	LmV	C	44.7	H = 03 35 51.9 h = 110 km MB = 5.3 (NEIC)
				D = 121.6
				LmH C 34s 0.5/um
				LmV C 25 0.4/um
15.	ePKIKP	A	06 29 27	Fiji Islands Region 20.61 S 178.83 W
	iPKHKP	AB	29 32.2	H = 06 10 50.9 h = 610 km MB = 5.5
	eiPKP2	A	29 37.5	D = 148.92 Az = 347.1 (NEIC)

Day	Phase		h m s	Remarks
cont. 15.	epPKP	A	06 31 50	PKIKPV A 2.1s 105.2nm
	e	A	31 53	PKHKPV A 1.6 308.0nm
				PKP2V A 1.2 97.5nm
15.	eP	A	13 23 36	Lake Tanganyika Region 7.18 S 30.27 E
	epP	A	23 44.5	H = 13 13 33.4 h = 36 km MB = 4.7
				D = 59.82 Az = 346.4 (NEIC)
				h = 34 km
				PV A 1.4s 23.3nm M = 5.1
15.	LmH	B	16 06.7	LmH B 19s 0.5µm
	LmV	B	06.7	LmV B 19 0.4µm
16.	eP	A	00 09 58	Crete 34.85 N 25.05 E
	LmH	B	16.5	H = 00 05 43.1 h = 43 km MB = 4.6
	LmV	B	18.8	D = 18.57 Az = 332.3 (NEIC)
				PV A 1.0s 11.8nm M = 4.0
				LmH B 15.5 0.9µm 4.2
				LmV B 11 0.5µm 4.3
16.	LmH	C	14 36.4	Greece 39.31 N 21.02 E
				H = 14 28 43.5 h = normal MB = 3.9 (NEIC)
				D = 13.13
				LmH C 16s 0.5µm M = 3.6
16.	+iP	A	14 59 57.5	Andreanof Islands, Aleutian Is.
	LmH	C	15 37.0	51.12 N 178.83 W
				H = 14 48 02.8 h = 54 km MB = 5.5
				D = 78.22 Az = 353.2 (NEIC)
				PV A 1.4s 58.1nm M = 5.4
				LmH C traces
17.	eP	A	03 45 56	Afghanistan 33.32 N 68.11 E
	e	A	46 00	H = 03 37 48.1 h = 40 km MB = 5.1
				D = 44.28 Az = 310.5 (NEIC)
				PV A 1.2s 16.3nm M = 4.7

Day	Phase		h m s	Remarks
17.	eP dif	A	12 49 19	<u>West New Guinea Region</u>
	ePKIKP	A	53 09	4.44 S 134.03 E
	eX	A	53 52	H = 12 34 26.5 h = normal MB=5.7 MS=6.4
	ePP	AB	54 08	D = 113.55 Az = 324.1 (NEIC)
	ePS	C	13 03 44	P difV A 1.8s 33.8nm
	eSS	C	09 15	XV A 2.1 76.7nm
	LmH	B	36.6	PPV A 1.8 138.4nm M = 5.4
	LmV	B	44.5	LmH B 20.5 20.2 μm 6.7
				LmV B 18 8.9 μm 6.4
17.	eP	A	14 35 43.5	<u>Philippine Islands Region</u>
				5.83 N 127.10 E
				H = 14 21 57.9 h = 69 km MB = 5.1
				D = 101.23 Az = 324.1 (NEIC)
				PV A 1.3s 13.1nm M = 5.4
17.	+eiP	AB	22 21 26.5	<u>Kurile Islands</u> 50.81 N 157.55 E
	eS	B	31 00	H = 22 09 49.1 h = 41 km MB=5.6 MS=5.4
	LmH	B	57.8	D = 74.84 Az = 338.3 (NEIC)
	LmV	B	58.0	PV A 1.8s 139.0nm M = 5.6
				LmH B 19 2.9 μm 5.6
				LmV B 20 2.7 μm 5.6
18.	ePKHKP	A	01 22 23.5	<u>Tonga Islands</u> 18.28 S 175.58 W
	ePKP2	A	23 38.5	H = 01 03 13.4 h = 290 km MB = 4.9
				D = 147.20 Az = 351.5 (NEIC)
				h = 283 km
				PKHKPV A 1.8s 27.0nm
18.	eP	A	03 19 22	<u>Central Mid-Atlantic Ridge</u>
	eS	C	27 34	4.00 N 31.51 W
	LmV	B	41.7	H = 03 09 23.3 h = normal MB = 4.8
	LmH	B	42.3	D = 58.87 Az = 30.6 (NEIC)
				PV A 1.6s 27.4nm M = 5.1
				LmV B 20 0.44 μm 4.6
18.	iPg	A	05 29 26	D c. 1.2
	eSg	A	29 42	

Day	Phase		h m s	Remarks
18.	ePKHKP	A	18 30 50.5	<u>Fiji Islands Region</u> 18.56 S 177.96 W
	ePKP2	A	30 53.5	H = 18 12 13.5 h = 617 km MB = 4.5
				D = 147.09 Az = 348.8 (NEIC)
19.	-iPKP	A	01 51 57.8	<u>Loyalty Islands</u> 20.19 S 168.81 E
				H = 01 32 24.4 h = 36 km MB=5.2 MS=5.0
				D = 144.67 Az = 334.7 (NEIC)
				PKPV A 1.1s 80.6nm
19.	LmV	C	02 49.2	<u>Either Loyalty Islands</u> 20.2 S 168.8 E
	LmH	C	49.5	H = 01 32 24.4 h = 36 km
				MB = 5.2 MS = 5.0 (NEIC)
				D = 144.7
				<u>Or Zambia</u> 16.8 S 28.2 E
				H = 02 12 12.7 h = normal MB = 4.6 (NEIC)
				D = 68.8
				LmV C 28s 0.6 μm M = 5.2 (Loy.Isl.)
				4.7 (Zambia)
19.	ePKHKP	A	03 36 04	<u>Fiji Islands Region</u> 21.26 S 178.73 W
				H = 03 17 17.5 h = 561 km MB = 4.3
				D = 149.56 Az = 347.0 (NEIC)
19.	+eP	A	04 40 43.5	<u>Eastern Kazakh SSR</u> 50.01 N 77.73 E
	ePn	A	42 15.5	H = 04 32 57.6 h = 0 km MB = 5.4
				D = 40.93 Az = 297.3 (NEIC)
				Underground explosion MB = 5.5 (UPP)
				PV A 0.9s 40.8nm M = 5.1
19.	e	A	04 42 40	<u>Near Coast of Peru</u> 14.68 S 76.32 W
				H = 04 28 45 h = 39 km MB = 4.6
				D = 99.9 Az = 40 (ISC)
19.	ePKHKP	AB	06 40 47	<u>Loyalty Islands Region</u> 22.33 S 170.32 E
	ePKP2	A	40 50.5	H = 06 21 04.9 h = 41 km MB=4.9 MS=5.1
	LmH	B	07 49.5	D = 147.19 Az = 334.7 (NEIC)

Day	Phase		h m s	Remarks
cont. 19.	LmV	B	07 49.5	PKHKPV A 1.9s 75.8nm
				PKHKPV B 8 1.3 μm
				LmH B 20 0.5 μm M = 5.2
				LmV B 20 0.6 μm 5.4
19.	e	A	07 09 32	LmH B 18s 0.45 μm
	LmH	B	14.2	LmV B 14 0.9 μm
	LmV	B	15.6	
19.	ePKP	A	11 45 35	<u>New Hebrides Islands</u> 18.33 S 169.43 E
				H = 11 26 36.4 h = 291 km MB = 4.6
				D = 143.22 Az = 336.3 (NEIC)
				PKPV A 1.0s 13.8nm
19.	ePn	A	17 43 25	<u>Central Italy</u> 43.47 N 12.47 E
	eiX	A	43 27.5	H = 17 41 37.2 h = 4 km MB = 4.4
	ePg	A	43 59	D = 7.20 Az = 355.7 (NEIC)
	e	B	45 10	PnV A 0.7s 17.2nm
	LmH	B	47.0	XV A 0.8 57.6nm
	LmV	B	47.1	LmH B 12.5 0.8 μm M = 3.6
				LmV B 12 0.8 μm
19.	eP	A	22 17 48.5	<u>Aegean Sea</u> 38.24 N 26.84 E
	LmH	B	23.5	H = 22 13 53.3 h = 13 km MB = 4.5
	LmV	B	24.4	D = 16.45 Az = 323.8 (NEIC)
				PV A 1.5s 32.7nm M = 4.2
				LmH B 13 1.4 μm 4.5
				LmV B 14 0.3 μm 3.8
20.	ePKP	A	00 52 40.5	<u>Samoa Islands Region</u> 14.83 S 173.14 W
				H = 00 33 10.1 h = 49 km MB = 5.3 MS=4.7
				D = 144.07 Az = 354.8 (NEIC)
				PKPV A 1.2s 30.3nm
20.	eP	A	09 32 00.5	<u>Near Coast of Guatemala</u>
				13.48 N 90.03 W
				H = 09 19 23.4 h = 88 km MB = 4.8
				D = 86.93 Az = 38.6 (NEIC)

Day	Phase		h m s	Remarks
20.	ePn	A	12 25 21.5	Gallery collapse in the Rochonvillers
	ePg	A	25 34.5	mine,(Luxembourg) 49.42 N 6.03 E
	eSn	A	26 09	H = 12 24 21 (BCIS)
	eiSg	A	26 29.5	h = 0 km MB = 4.3
				D = 3.80 Az = 69.0 (NEIC)
20.	iPn	A	14 00 30.8	Czechoslovakia 50.59 N 14.05 E
	ePg	A	00 32	explosion yield 15 t (PRU)
	eSn	A	00 51	H = 14 00 02
	eiSg	A	00 52.4	D = 1.55 Az = 273 (ISC)
20.	ePKHKP	A	18 40 48.5	Tonga Islands 18.59 S 175.51 W
				H = 18 21 22.1 h = 159 km MB = 4.7
				D = 147.52 Az = 351.5 (NEIC)
21.	ePKIKP	A	05 29 47	South of Kermadec Islands
	e	A	29 54	33.19 S 179.14 W
	e	A	30 30.5	H = 05 09 48.6 h = normal MB = 5.3
				D = 160.85 Az = 338.8 (NEIC)
21.	LmH	B	15 39.0	LmH B 20s 0.3/um
	LmV	B	47.5	LmV B 20 0.4/um
21.	ePKP2	A	17 25 50	Probably: Kermadec Islands
				30 S 177 W
				H = 17 05 18 MB = 4.4 (NORSAR)
				D = 158.2
21.	LmV	C	20 48.5	LmV C 16s 0.35/um
	LmH	C	49	
21.	ePKIKP	A	21 28 42	New Guinea 6.42 S 144.25 E
				H = 21 09 51.2 h = 29 km MB = 5.2
				D = 120.94 Az = 326.9 (NEIC)
21.	eP	A	22 23 33	Near East Coast of Kamchatka
				55.11 N 161.90 E
				H = 22 12 12.4 h = normal MB = 4.4
				D = 71.73 Az = 340.6 (NEIC)

Day	Phase		h m s	Remarks
21.	LmH	B	22 27.3	LmH B 17s 0.25 /um
	LmV	B	27.3	LmV B 17 0.15 /um
22.	eX	A	05 57 42	Yunnan Province, China 27.67 N 104.07 E
	LmH	B	06 25.0	H = 05 46 20.9 h = normal MB=5.0 MS=4.9
	LmV	B	34.5	D = 70.61 Az = 317.6 (NEIC)
				XV A 1.6s 22.0nm
				LmH B 22 0.7 /um M = 4.9
				LmV B 15 0.25 /um 4.6
22.	LmH	B	12 42.8	Ryukyu Islands 29.45 N 129.89 E
	LmV	B	43.0	H = 11 49 08.7 h = 48 km MB = 4.8 (NEIC)
				D = 83.5
				LmH B 18s 0.2 /um M = 4.6
				LmV B 16 0.3 /um 4.8
22.	eP	A	13 43 51	Crete 35.18 N 23.40 E
	e	A	43 53.5	H = 13 39 46.1 h = 56 km MB = 4.5
	epP	A	44 03	D = 17.67 Az = 334.6 (NEIC)
	LmH	B	52.2	PV A 0.8s 17.3nm M = 4.3
	LmV	B	52.2	LmH B 13.5 0.5 /um
				LmV B 15 0.4 /um
22.	eP	A	21 36 42	Western Iran 30.74 N 49.83 E
	LmH	B	52.7	H = 21 29 57.2 h = 57 km MB = 5.0
	LmV	B	54.3	D = 34.62 Az = 316.1 (NEIC)
				PV A 1.5s 15.1nm M = 4.7
				LmH B 16 0.4 /um
				LmV B 14 0.4 /um
22.	eP	A	21 54 11	Andreanof Islands, Aleutian Is.
				51.13 N 179.84 W
				H = 21 42 16.1 h = 54 km MB = 4.8
				D = 78.14 Az = 352.6 (NEIC)
23.	eP	A	03 01 43	Iceland 64.63 N 17.55 W
				H = 02 57 00.2 h = 5 km MB = 4.5
				D = 20.70 Az = 118.6 (NEIC)
				PV A 1.8s 20.1nm M = 4.2

Day	Phase		h m s	Remarks
23.	ePn	A	13 40 10.5	Northern Italy 44.05 N 12.77 E
	e	A	40 12	H = 13 38 34.0 h = normal MB = 4.2
	e	A	40 17	D = 6.65 Az = 353.6 (NEIC)
	ePg	A	40 50	PnV A 0.9s 31.1nm
	eSn	A	41 23	LmH B 12 0.9 /um M = 3.5
	eSg	A	42 11	LmV B 12 1.2 /um
	LmH	B	43.4	
	LmV	B	43.6	
23.	LmH	B	14 23.7	Taiwan 24.01 N 121.51 E
	LmV	B	23.9	H = 13 29 16.0 h = 42 km MB = 5.2 (NEIC)
				D = 83.4
				LmH B 17.5s 0.7 /um M = 5.1
				LmV B 14 0.6 /um 5.1
23.	ePKIKP	A	14 31 16	South of Kermadec Islands
	ePKP2	A	32 02.5	34.51 S 179.81 E
				H = 14 11 19.2 h = normal MB = 5.3 (NEIC)
				D = 161.72
				PKIKPV A 1.6s 13.7nm
23.	ePKHKP	A	18 48 23	Fiji Islands Region 20.90 S 179.29 W
	ePKP2	A	48 29.5	H = 18 29 44.8 h = 645 km MB = 4.0
				D = 149.09 Az = 346.4 (NEIC)
24.	eP	A	04 02 47	Baffin Island Region 71.73 N 71.27 W
				H = 03 55 06.8 h = normal MB = 4.2
				D = 40.83 Az = 75.1 (NEIC)
24.	LmH	B	04 56.8	Philippine Islands 7.09 N 126.99 E
	LmV	B	05 08.5	H = 03 52 33.6 h = normal MB=5.0
				MS = 4.8 (NEIC)
				D = 100.2
				LmH B 16s 0.4 /um M = 5.0
				LmV B 17 0.5 /um 5.1
24.	eP	A	10 02 00	Off East Coast of Honshu, Japan
	eS	C	12 00	40.63 N 143.52 E

Day	Phase		h m s	Remarks
cont. 24.	LmH	B	10 35.9	H = 09 49 53.1 h = 32 km MB = 4.9
	LmV	B	43.3	D = 79.83 Az = 331.2 (NEIC)
				PV A 1.2s 10.2nm M = 4.7
				SH C traces
				LmH B 17 1.3/um 5.4
				LmV B 17 0.7/um 5.1
24.	eSg	A	16 22 34	<u>Poland</u> 50.23 N 19.32 E
				H = 16 19 55.7 M = 3.0 (WAR)
				D = 4.6
24.	. eP	AB	18 54 54	<u>Colombia</u> 5.16 N 75.21 W
	epP	AB	55 23	H = 18 42 31.9 h = 118 km MB = 5.5
	esP	A	55 40	D = 84.41 Az = 39.7 (NEIC)
	eS	B	19 05 05	h = 125 km
	ePS	C	05 52	PV A 1.4s 14nm M = 4.7
	eSS	C	10 40	LmH, LmV traces
	LmH	B	20.5	
	LmV	B	25.5	
24.	eP	AB	21 29 31.5	<u>South of Panama</u> 7.27 N 82.34 W
	eS	B	40 05	H = 21 16 45.4 h = normal MB = 5.5
				D = 86.94 Az = 39.5 (NEIC)
				PV A 2.2s 87.0nm M = 5.7
24.	-iP	AB	21 42 47.0	<u>South of Panama</u> 4.96 N 78.14 W
	ei	B	43 07	H = 21 30 09.9 h = 50 km MB=6.3 MS=6.5
	ei	B	43 29	D = 86.05 Az = 39.7 (NEIC)
	ePP	B	46 00	PV A 1.9s 743.0nm M = 6.6
	+iS	B	53 22	PV B 8.5 9.4/um 7.1
	ePS	B	54 16	PPV B 9 5.7/um 7.0
	eiSS	C	59 00	SH B 15 17.6/um 7.0
	ePKKP	A	22 00 50	LmH B 18 9.3/um 6.2
	eP'P'	A	08 48	LmV B 17 9.2/um 6.3
	e	A	09 06	P'P'P'V A 3.5 291.7nm
	eiLR	C	10 45	
	e	A	13 47	
	LmH	B	28.9	
	LmV	B	29.0	
	eP'P'P'	A	29 39	

Day	Phase		h m s	Remarks
24.	eP	A	22 59 42	South of Panama 4.86 N 78.18 W H = 22 47 01.8 h = normal MB = 5.3 D = 86.15 Az = 39.7 (NEIC) PV A 1.4s 32.6nm M = 5.4
25.	eP	A	02 47 30.5	South of Panama 4.75 N 78.05 W H = 02 34 48.7 h = normal MB = 4.9 D = 86.15 Az = 39.6 (NEIC)
25.	-iP	AB	03 24 49	Tadzhik SSR 37.60 N 72.11 E
	e	A	25 39	H = 03 16 52.3 h = 136 km MB = 5.2
	ePP	A	26 37	D = 44.14 Az = 307.3 (NEIC)
	e	A	27 41	PV A 2.0s 120.0nm M = 5.2
	LmV	B	45.5	LmH B 11 0.4/um
	LmH	B	45.6	LmV B 13 0.4/um
25.	ePKP	A	06 59 23.5	Tonga Islands 15.61 S 175.82 W H = 06 40 27.7 h = 350 km MB = 5.1 D = 144.54 Az = 351.8 (NEIC) PKPV A 1.2s 34.6nm
25.	eP	A	07 58 19	South of Panama 4.95 N 78.15 W H = 07 45 38.7 h = normal MB = 4.8 D = 86.06 Az = 39.7 (NEIC)
25.	ePKHKP	A	08 22 29.5	Tonga Islands 17.82 S 173.25 W
	epPKHKP	A	22 55	H = 08 02 55.1 h = 78 km MB = 4.6 D = 147.02 Az = 354.3 (NEIC)
25.	eP	A	08 43 17	Southern Iran 26.84 N 55.44 E
	LmV	B	09 15.7	H = 08 35 37.4 h = 43 km MB = 4.6
	LmH	B	15.8	D = 40.83 Az = 317.6 (NEIC) PV A 1.9s 22.7nm M = 4.6 LmH B 18 0.5/um 4.4 LmV B 19 0.7/um 4.6
25.	eP	AB	14 33 42	South of Honshu, Japan
	eS	B	44 00	33.39 N 140.75 E

Day	Phase		h m s	Remarks
cont. 25.	LmH	B	15 12.7	H = 14 21 13.2 h = 65 km MB = 5.7
	LmV	B	18.5	D = 85.05 Az = 330.3 (NEIC)
				PV A 1.5s 93.0nm M = 5.6
				LmH B 16 1.3 µm
				LmV B 18 1.3 µm
25.	e	A	17 35 06	<u>Volcano Islands Region</u>
	e	A	35 17	23.51 N 143.22 E
	ePP	A	38 48	H = 17 21 39.5 h = normal MB=5.0 MS=4.8
	e	A	39 02	D = 94.72 Az = 331.5 (NEIC)
	LmH	B	18 21.0	First onset traces
	LmV	B	21.3	LmH B 19.5s 0.7 µm M = 5.1
				LmV B 19 0.6 µm 5.1
25.	eP	A	19 12 30.5	<u>South of Panama</u> 4.60 N 78.31 W
				H = 18 59 48.7 h = normal MB = 4.8
				D = 85.43 Az = 39.6 (NEIC)
				PV A 1.2s 12.2nm M = 5.0
25	+iPKP2	A	20 33 33.5	<u>Kermadec Islands Region</u> 31.67 S 179.70 E
				H = 20 13 41.4 h = 387 km MB= 4.7 (NEIC)
				D = 159.07
				PKP2V A 1.4s 60.5nm
25.	eP dif	C	21 49 15	<u>South Sandwich Islands Region</u>
	ePKIKP	A	53 09	59.28 S 26.10 W
	e	C	54 44	H = 21 34 37.9 h = 67 km MB = 6.0
	e(PKS)	B	56 45	D = 113.78 Az = 25.2 (NEIC)
	ePS	B	22 03 38	PKIKPV A 1.3s 13.1nm
	ePKKP	A	04 13	LmH A 23 4.6 µm
	e	A	04 18	LmV A 22 4.2 µm
	ePPS	C	05 20	
	eSS	C	10 05	
	eSSS	C	13 57	
	LmH	B	35.2	
	LmV	B	35.3	

Day	Phase		h m s	Remarks
26.	eP	A	02 13 07	<u>Hokkaido, Japan Region</u> 41.17 N 142.77 E
				H = 02 01 04.7 h = 35 km MB = 4.2
				D = 79.08 Az = 330.8 (NEIC)
				traces
26.	eP	A	07 55 39	<u>South of Panama</u> 5.08 N 78.2 W
	e	A	55 54	H = 07 42 59.5 h = normal MB=5.2 MS=4.3
				D = 85.97 Az = 39.7 (NEIC)
				PV A 2.0s 51.3nm M = 5.4
26.	e(pP)	A	14 38 19	<u>Southern Iran</u> 27.09 N 60.84 E
	ePP	A	39 55.5	H = 14 30 05.4 h = 46 km MB = 5.2
				D = 43.94 Az = 316.0 (NEIC)
				PPV A 1.3s 8.7nm M = 4.4
26.	i	A	15 00 03.5	Explosion yield 20 t (PRU)
				49.26 N 14.92 E
				D = 2.5
26.	eP	A	17 27 18.5	<u>Southern Nevada</u> 37.12 N 116.06 W
				H = 17 15 00.2 h = 0 km MB = 5.6
				D = 81.23 Az = 30.7 (NEIC)
				PV A 1.2s 32.5nm M = 5.3
				Nuclear explosion (AEC)
26	eP dif	A	20 40 55	<u>Hawaii</u> 19.93 N 155.10 W
	ePKIKP	A	45 02	H = 20 26 28.0 h = 50 km MB=6.0 MS=6.1
	e	A	45 10	D = 108.76 Az = 8.9 (NEIC)
	ePP	AC	45 25	P dif A 2.2s 38.2nm
	eSKS	C	51 34	PPV C 23 1.5/um M = 6.3
	eS dif	C	52 55	LmH B 18.5 3.5/um 6.0
	ePS	C	54 52	LmV B 19 3.9/um 6.0
	ePKKP	A	56 07	
	eSS	C	21 00 33	
	eP'P'	A	04 34	
	LmV	B	38.6	
	LmH	B	38.7	

Day	Phase		h m s	Remarks
27.	eP	A	00 35 27.5	Turkey 38.71 N 33.00 E
	eS	C	39 05	H = 00 31 03.0 h = normal MB = 4.6
	LmH	B	42.3	D = 19.25 Az = 315.2 (NEIC)
	LmV	B	44.5	PV A 2.0s 51.3nm M = 4.4
				LmH B 18 0.7 μm 4.0
				LmV B 14 0.8 μm 4.3
27.	ePP	AC	18 36 12	Celebes 2.84 S 119.48 E
	eSS	C	50 30	H = 18 17 49.4 h = 33 km MB = 5.1 (NEIC)
	LmH	B	19 25.3	D = 103.47
	LmV	B	25.4	LmH B 20s 0.9 μm M = 5.3
				LmV B 20 0.9 μm 5.3
27.	e	A	21 46 11.5	
28.	eP	A	03 02 49	Crete 34.25 N 26.32 E
				H = 02 58 27.6 h = normal MB = 4.4
				D = 19.59 Az = 331.2 (NEIC)
28.	eP	A	04 55 33	Southwestern Ryukyu Islands
	LmH	B	05 37.5	25.67 N 125.09 E
	LmV	B	37.5	H = 04 43 11.4 h = 103 km MB = 4.9
				D = 84.07 Az = 324.0 (NEIC)
				PV A 1.6s 11.0nm M = 4.5
				LmH B 16 2.06 μm
				LmV B 16 0.82 μm
28.	eP	A	11 02 37	Southern Sumatra 4.04 S 102.60 E
	epP	A	02 55	H = 10 49 28.2 h = 82 km MB = 5.4
				D = 93.72 Az = 320.4 (NEIC)
				h = 68 km
28.	eP	A	12 34(10)	Ecuador 1.44 S 79.88 W
	e	A	34 17	H = 12 21 11.7 h = 109 km MB = 5.5
	epP	A	34 37.5	D = 92.05 Az = 39.6 (NEIC)
				h = 103 km
				PV A traces

Day	Phase		h m s	Remarks
28.	ePKP	A	21 32 58	Fiji Islands Region 17.35 S 178.52 W H = 21 14 06.7 h = 398 km MB = 4.5 D = 145.80 Az = 348.5 (NEIC) PKPV A 1.2s 16.2nm
28.	LmH LmV	C C	21 38.0 38.7	Borneo 4.6 N 117.7 E H = 20 39 43.9 h = normal MB = 5.4 (NEIC) D = 96.5
29.	eP1 eP2 LmH LmV	A AB B B	14 42 25 42 26.5 50.2 50.7	Morocco 34.64 N 4.07 W H = 14 37 53.9 h = 19 km MB = 4.6 D = 19.66 Az = 30.8 (NEIC) P1V A 1.0s 15.7nm M = 4.2 P2V A 1.4 55.8nm 4.6 P2H A 1.6 70.0nm 4.8 LmH B 12 0.9 μm 4.3 LmV B 11 0.4 μm 4.2
29.	ePKHKP ePKP2	A A	18 56 30.8 56 34.5	Fiji Islands Region 19.83 S 177.66 W H = 18 37 25.5 h = 364 km MB = 4.9 D = 148.38 Az = 348.7 (NEIC)
29.	eP e e LmH LmV	A A A B B	20 04 48 04 55.5 05 10 28.8 32.8	West Pakistan 29.30 N 68.69 E H = 19 56 14.3 h = 24 km MB = 4.9 D = 47.33 Az = 313.4 (NEIC) PV A 1.2s 16.3nm M = 4.9 LmH B 15 1.0 μm 4.9 LmV B 12 1.0 μm 5.1
29.	eP	A	21 47 31.5	Near East Coast of Kamchatka 56.84 N 161.71 E H = 21 36 19.5 h = normal MB = 4.8 D = 70.07 Az = 340.3 (NEIC) PV A 1.5s 17.6nm M = 4.9
29.	ePKIKP ePKP2	A A	23 50 27 50 35	Fiji Region 20.12 S 178.68 W H = 23 30 44.1 h = 0 km MB = 4.9 D = 148.46 Az = 347 (ISC)

Day	Phase		h m s	Remarks
30.	ePKHKP	A	02 13 02	South of Fiji Islands
	ePKP2	A	13 14.5	25.39 S 179.89 E
				H = 01 53 57.3 h = 479 km MB = 4.7
				D = 153.21 Az = 343.3 (NEIC)
30.	eP	A	07 38 19.5	Central Russia 50.99 N 89.73 E
	e	A	38 24	H = 07 29 48.7 h = normal MB = 4.6
				D = 47.12 Az = 301.8 (NEIC)
				PV A 1.2s 12.2nm M = 4.8
30.	ePKIKP	A	08 57 39	Fiji Islands 17.51 S 179.62 E
	iPKHKP	A	57 40.5	H = 08 39 07.7 h = 613 km MB = 5.4
	iPKP2	A	57 42.5	D = 145.58 Az = 346.5 (NEIC)
	iSKP	A	09 00 24	PKHKPV A 1.6s 68.7nm
				SKPV A 1.0 19.7nm
30.	eP	A	15 53 27	Western Caucasus 43.34 N 43.77 E
				H = 15 48 24.8 h = 46 km MB = 4.2
				D = 22.98 Az = 299.8 (NEIC)
				traces
30.	e	A	21 13 29.5	South of Panama 4.99 N 77.96 W
				H = 21 00 41.0 h = 23 km MB = 5.0
				D = 85.91 Az = 39.7 (NEIC)
30.	eP	A	22 50 08.5	Near East Coast of Honshu, Japan
	LmH	B	23 31.4	36.23 N 141.42 E
	LmV	B	32.2	H = 22 37 47.9 h = 51 km MB = 4.8
				D = 82.87 Az = 330.5 (NEIC)
				LmH B 16s 0.6 µm
				LmV B 14 0.4 µm

Day	Phase		h m s	Remarks
1.	-eP	A	01 04 36.5	Kurile Islands Region 43.94 N 148.43 E
	epP	A	04 51	H = 00 52 39.7 h = 57 km MB = 4.9
				D = 78.58 Az = 333.6 (NEIC)
				h = 55 km
				PV A 1.0s 23.6nm M = 5.1
1.	ePn	A	05 46 34.5	Adriatic Sea 42.62 N 17.15 E
	eSn	A	48 03.5	H = 05 44 25.1 h = normal MB = 3.8
				D = 8.88 Az = 336.6 (NEIC)
1.	ePKIKP	A	08 01 43	Fiji Islands Region 17.96 S 178.31 W
	ePKHKP	AB	01 45.5	H = 07 43 09.1 h = 596 km MB = 5.0
	ePKP2	A	01 48	D = 146.44 Az = 348.6 (NEIC)
				PKIKPV A 1.4s 30.2nm
				PKHKPV A 1.8 152.0nm
1.	ePKHKP	A	10 13 19	Tonga Islands 19.81 S 173.83 W
				H = 09 53 31.7 h = normal MB = 5.0
				D = 148.93 Az = 353.3 (NEIC)
1.	ePKIKP	A	10 59 50.5	East New Guinea Region 10.02 S 150.18 E
	ePP	AB	11 01 49	H = 10 40 46.9 h = 27 km MB=5.9 MS=5.4
	ePS	C	11 55	D = 127.11 Az = 328.1 (NEIC)
	eSS	C	18 50	PKIKPV A 1.8s 60.8nm
	LmH	B	46.5	LmH B 23 3.0 μm M = 5.9
	LmV	B	12 01.0	LmV B 18 1.6 μm 5.7
1.	ePKHKP	A	15 34 12	South of Fiji Islands 25.16 S 179.39 E
	ePKP2	A	34 26	H = 15 15 15.2 h = 544 km MB = 4.7
	e	A	34 44.5	D = 152.86 Az = 342.8 (NEIC)
				PKHKPV A 1.5s 25.1nm
2.	ePKIKP	A	01 45 23.5	East New Guinea Region 9.98 S 150.18 E
	e	A	45 31	H = 01 26 20.9 h = 29 km MB=5.6 MS=5.2
	ePP	AB	47 24	D = 127.09 Az = 328.1 (NEIC)
	ePS	C	57 28	PKIKPV A 2.0s 51.3nm
	eSS	C	02 04 25	LmH B 18 1.1 μm M = 5.6
	LmH	B	46.8	LmV B 18 0.9 μm 5.5
	LmV	B	47.0	

Day	Phase		h m s	Remarks
2.	eP	A	05 31 00.5	Kurile Islands 44.00 N 148.36 E
				H = 05 19 02.9 h = 48 km MB = 4.5
				D = 78.51 Az = 333.6 (NEIC)
				PV A 1.1s 10.1nm M = 4.7
2.	-ePKP	A	13 58 41	Near Coast of Southern Chile
	LmH	C	14 37.8	48.91 S 75.84 W
				H = 13 39 45.4 h = normal MB = 5.2
				D = 124.11 Az = 50.2 (NEIC)
2.	e	A	16 46 16.5	
2.	eP1	A	23 22 08	North Atlantic Ocean 36.47 N 12.33 W
	iP2	A	22 10	H = 23 17 11.6 h = normal MB = 4,4 (NEIC)
	LmH	B	30.0	D = 22.27
	LmV	B	31.7	P2V A 1.5s 45.2nm M = 4.7
				LmH B 18 0.5/um 4.0
				LmV B 16 0.2/um 3.9
3.	eP	A	01 25 49.5	Ionian Sea 38.11 N 19.77 E
				H = 01 22 33.3 h = 32 km MB = 4.0
				D = 13.81 Az = 337.8 (NEIC)
3.	ePKP	A	04 56 03.5	Fiji Islands Region 16.27 S 176.61 W
				H = 04 37 15.0 h = 429 km MB = 4.3
				D = 145.08 Az = 350.8 (NEIC)
3.	eP	A	05 51 09	Queen Elizabeth Islands
	e	A	51 22	76.60 N 107.41 W
	LmH	B	06 11.5	H = 05 42 34.1 h = normal MB = 4.3
	LmV	B	13.6	D = 47.32 Az = 49.2 (NEIC)
				LmH B 14s 0.8/um M = 4.8
				LmV B 16 0.5/um 4.6
3.	eP	A	07 51 41	Southern Iran 28.20 N 51.90 E
				H = 07 44 24.4 h = 41 km MB = 4.6
				D = 37.71 Az = 317.7 (NEIC)

Day	Phase		h m s	Remarks
3.	ePKP	A	13 45 06	<u>Fiji Islands Region</u> 17.93 S 178.41 W H = 13 26 31.0 h = 600 km MB = 5.1 D = 146.39 Az = 348.5 (NEIC) PKPV A 1.3s 21.8nm
3.	e	A	20 23 22	
3.	e	A	22 05 30.5	
3.	ePP	AB	23 30 12	<u>Kerguelen Islands Region</u>
	ePS	C	39 36	46.10 S 73.20 E
	eSS	C	45 42	H = 23 11 05.7 h = 33 km MB = 5.5 (NEIC)
	LmV	B	24 24.0	D = 110.06
	LmH	B	26.5	PPV A 2.7s 134.0nm M = 6.2
				LmH B 17 0.9 μm 5.4
				LmV B 18 0.9 μm 5.4
4.	ePKP	A	01 39 57	<u>Fiji Islands Region</u> 17.59 S 178.60 W H = 01 21 21.7 h = 590 km MB = 4.0 D = 146.03 Az = 348.3 (NEIC)
4.	eP	A	11 41 15.5	<u>Molucca Passage</u> 2.28 N 126.72 E
	ePP	A	45 32	H = 11 27 13.4 h = normal MB=5.9 MS=5.5
	eSKS	B	51 51	D = 103.85 Az = 323.6 (NEIC)
	eS	C	53 00	PV A 1.5s 45.2nm M = 6.1
	eSP	B	54 24	PPV A 1.6 38.5nm 5.7
	eSS	C	12 00 00	LmH B 19.5 2.7 μm 5.8
	LmH	B	26.4	LmV B 15.5 1.5 μm 5.6
	LmV	B	33.2	
4.	ePn	A	13 01 26	<u>Czechoslovakia</u> 50.62 N 14.35 E Explosion, 7.6 t H = 13 00.9 (PRU) D = 1.74 Az = 722
4.	eP	A	14 10 32	<u>North Atlantic Ridge</u> 31.08 N 41.46 W H = 14 02 29.8 h = 33 km MB = 4.5 D = 43.59 Az = 48 (ISC)

Day	Phase		h m s	Remarks
4.	-eiP	AB	24 09 02.5	Southern Sumatra 1.47 S 99.85 E
	e	A	09 13	H = 23 56 06.1 h = 51 km MB = 5.9
	eSKS	C	19 24	D = 90.00 Az = 320.5 (NEIC)
	eS	C	19 50	PV A 1.3s 87.4nm M = 5.7
	ePS	C	20 52	LmH B 22 1.3/um
	eSS	C	25.6	LmV B 20 1.0/um
	eSSS	C	29.3	
	LmH	B	54.7	
	LmV	B	54.9	
5.	ePKIKP	A	01 54 29	Solomon Islands 8.15 S 156.41 E
				H = 01 35 19.5 h = 15 km MB = 5.4
				D = 128.58 Az = 332.0 (NEIC)
5.	eP	A	04 04 44	Near East Coast of Honshu, Japan
	epP	A	04 55.5	37.13 N 141.32 E
	e	A	05 34	H = 03 52 26.4 h = 41 km MB = 5.4
	LmH	B	44.3	D = 82.05 Az = 330.4 (NEIC)
	LmV	B	45.7	h = 44 km
				PV A 1.4s 51.1nm M = 5.4
				LmH B 16 0.3/um 4.8
				LmV B 18 0.4/um 4.8
5.	eP	A	05 24 37.5	Near East Coast of Honshu, Japan
				36.15 N 140.06 E
				H = 05 12 23.3 h = 80 km MB = 4.8
				D = 82.38 Az = 329.8 (NEIC)
				PV A traces
5.	e	A	06 19 56	Iran 33.34 N 57.38 E
	LmH	B	36.1	H = 06 12 35.6 h = normal MB = 4.6
	LmV	B	37.2	D = 37.48 Az = 311.4 (NEIC)
5.	e(PKP)	A	10 27 51	South of Fiji Islands 24.68 S 179.48 E
				H = 10 09 04.2 h = 586 km
				D = 152.43 Az = 343.2 (NEIC)
				PKPV traces

Day	Phase		h m s	Remarks
5.	ePKP	A	19 12 56	Fiji Islands Region 16.63 S 176.96 W
	eX1	A	12 59.5	H = 18 53 20.3 h = normal MB=5.3 MS=5.5
	eX2	B	13 15	D = 145.37 Az = 350.4 (NEIC)
	LmH	B	20 12.5	X1V A 1.5s 45.2nm
	LmV	B	16.0	X2V A 5.5 0.7/um
				LmH B 22 0.9/um M = 5.5
				LmV B 20 0.4/um 5.2
6.	eP	A	04 06 57.5	Southern Iran 27.30 N 55.52 E
	e	A	07 13.5	H = 03 59 19.6 h = 28 km MB = 4.9
				D = 40.54 Az = 317.2 (NEIC)
				Central Iran 27.5 N 55.3 E
				H = 03 59 29 h = 75 km (BCIS)
6.	ePn	A	04 58 44	Austria 47.4 N 15 E
	eSn	A	59 32	H = 04 57 45 (BCIS)
	iSg	A	59 46	D = 3.70 Az = 326 (ISC)
6.	ePn	A	08 19 12.5	Swabian Region 48.3 N 9.0 E
	ePg	A	19 21	H = 08 18 29 (BCIS)
	eSg	A	20 00	D = 2.89
6.	eiP	AB	14 50 07.5	Near E. Coast of Eastern Russia
	epP	A	51 55	43.52 N 132.26 E
	eS	B	58 52	H = 14 39 28.1 h = 497 km MB = 5.3
	eSP	AB	59 20	D = 72.96 Az = 325.0 (NEIC)
	LmH	B	15 09.9	h = 513 km
				PV A 1.8s 108.0nm M = 5.1
				LmH B 18 0.5/um
6.	eP	A	15 23 00.5	Near East Coast of Kamchatka
				53.93 N 160.49 E
				H = 15 11 37.8 h = 57 km MB = 5.0
				D = 72.57 Az = 339.8 (NEIC)
				PV A 1.5s 42.4nm M = 5.2
6.	LmH	B	19 10.8	LmH B 16s 0.3/um
	LmV	B	12.9	LmV B 12 0.3/um

Day	Phase		h m s	Remarks
6.	LmH	B	24 09.2	<u>Northern Easter I. Cordillera</u>
	LmV	B	14.8	4.43 S 105.01 W
				H = 23 03 47.5 h = normal MB = 4.5 (NEIC)
				D = 110.0
7.	eP	A	11 05 26	<u>Nicobar Islands Region</u> 7.16 N 91.30 E
	e	A	05 43	H = 10 53 29.2 h = 35 km MB = 4.8
	e	A	05 51	D = 77.96 Az = 320.2 (NEIC)
	LmH	C	40	PV A 1.5s 15.1nm M = 4.8
	LmV	C	47.5	
7.	eP	A	15 01 36.5	<u>South of Panama</u> 4.57 N 78.31 W
	e	A	01 40.5	H = 14 48 54.6 h = normal MB = 4.9
				D = 86.45 Az = 39.6 (NEIC)
7.	LmV	B	17 26.3	<u>East Central Pacific Ocean</u>
	LmH	B	26.4	16.57 N 116.08 W
				H = 16 27 12.1 h = normal MB = 4.8 (NEIC)
				D = 98.9
				LmH B 16s 0.5/um M = 5.1
				LmV B 16 0.5/um 5.1
7.	LmH	B	20 22.1	LmH B 16s 0.3/um
	LmV	B	24.2	LmV B 16 0.2/um
7.	e	A	20,29 19	
8.	ePKHKP	A	01 02 56	<u>New Hebrides Islands</u> 15.72 S 167.83 E
	ePKIKP	A	03 00	H = 00 43 47.9 h = 153 km MB = 5.2
	e	A	03.02	D = 140.23 Az = 336.3 (NEIC)
				PKHKPV traces
8.	ePKIKP	A	05 03 35	<u>Fiji Islands Region</u> 17.60 S 178.93 W
	ePKHKP	A	03 37	H = 04 44 56.5 h = 543 km MB = 5.3
	ePKP2	A	03 39.5	D = 145.97 Az = 348.0 (NEIC)
				PKHKPV A 1.5s 161.0nm

Day	Phase		h m s	Remarks
8.	eP	A	08 00 46	Kurile Islands 45.65 N 149.56 E
	e	A	00 51	H = 07 48 59.8 h = 95 km MB = 5.4
				D = 77.41 Az = 334.1 (NEIC)
				PV A 1.7s 60.6nm M = 5.2
8.	eP	A	10 12 29	Kansu Province, China 32.99 N 103.97 E
	LmH	B	43.1	H = 10 01 38.2 h = normal MB = 4.7
	LmV	B	43.2	D = 66.68 Az = 316.2 (NEIC)
				PV A 1.5s 15.1nm M = 4.9
				LmH B 15 0.4/um 4.8
				LmV B 16 0.5/um 4.9
8.	ePn	A	19 09 39	Northern Italy 45.66 N 9.71 E
	ePg	A	10 03	H = 19 08 24.0 h = normal
	eSn	A	10 35.5	D = 5.15 Az = 13.7 (NEIC)
	eSg	A	11 11	
9.	e	A	13 22 35	
10.	eP	A	11 51 24	Andreanof Islands, Aleutian Is.
				51.37 N 179.52 W
				H = 11 39 31.5 h = 61 km MB = 5.3
				D = 77.92 Az = 352.8 (NEIC)
				PV A 1.2s 44.7nm M = 5.3
10.	e(P)	A	16 31 06	
	e(Sg)	A	31 42	
10.	eP	A	18 03 55	Near Coast of Jalisco, Mexico
	LmH	B	45.0	19.01 N 104.76 W
	LmV	B	52.6	H = 17 50 53.1 h = normal MB = 5.0
				D = 91.03 Az = 34.8 (NEIC)
				PV A 2.0s 34.2nm M = 5.4
				LmH B 16 0.4/um 5.0
				LmV B 16 0.5/um 5.0

Day	Phase		h m s	Remarks
10.	eP	A	23 58 17.5	Mid-Indian Rise 12.20 S 65.77 E
				H = 23 46 19.8 h = normal MB = 4.8
				D = 78.31 Az = 328.2 (NEIC)
				PV A 2.0s 25.6nm M = 4.9
11.	eP	A	00 14 20.5	Greenland Sea 79.40 N 3.08 E
				H = 00 08 22.0 h = normal MB = 5.0
				D = 29.03 Az = 168.8 (NEIC)
				PV A 1.1s 16.1nm M = 4.6
11.	ePKHKP	A	05 31 33.5	Tonga 21.59 S 175.76 W
	e	A	32 11	H = 05 11 54 h = 112 km MB = 4.7
				D = 150.43 Az = 350 (ISC)
11.	eP	A	10 59 33	Molucca Passage 0.98 N 126.03 E
	eS	B	11 11 20	H = 10 45 25.4 h = 24 km MB=5.4 MS=5.2
	LmH	B	48.5	D = 104.48 Az = 323.2 (NEIC)
	LmV	B	50.8	PV A 1.8s 27.0nm M = 5.8
				LmH B 22 1.1/um 5.4
				LmV B 20 1.2/um 5.4
11.	eP	A	13 59 44.5	Iran 33.35 N 57.38 E
	eS	B	14 05 34	H = 13 52 31.7 h = 50 km MB = 5.1
	LmH	B	16.3	D = 37.46 Az = 311.4 (NEIC)
	LmV	B	18.4	PV A 2.0s 51.3nm M = 5.1
				LmH B 20 2.4/um 5.0
				LmV B 15 1.9/um 5.1
11.	ePKP2	A	16 16 20.5	Kermadec Islands Region 31 S 180 W
				H = 15 55 51 (NORSAR)
				D = 158.7
12.	eP	A	05 59 55	Southern Sumatra 3.24 S 101.24 E
	epP	A	06 00 10	H = 05 46 48.7 h = 51 km MB = 5.5
	e	A	00 20	D = 92.24 Az = 320.4 (NEIC)
				h = 59 km
				PV A 1.4s 18.6nm M = 5.3

Day	Phase		h m s	Remarks
12.	eP	A	07 17 03	Unimak Island Region 53.57 N 163.72 W
				H = 07 05 16.6 h = normal MB = 4.3
				D = 76.09 Az = 3.1 (NEIC)
				PV A 1.0s 11.8nm M = 4.8
12.	eP	A	07 35 35.5	Unimak Island Region 53.54 N 163.69 W
	epP	A	35 47	H = 07 23 49.4 h = 30 km MB = 4.5
				D = 76.11 Az = 3.1 (NEIC)
				h = 41 km
				PV A 1.2s 20.3nm M = 5.0
12.	ePKIKP	A	16 39 07	New Ireland Region 3.73 S 152.08 E
	e	A	39 10	H = 16 20 09.2 h = 13 km MB=5.5 MS=5.9
	ePP	AB	40 47	D = 122.67 Az = 331.2 (NEIC)
	e	B	53 10	LmH B 19s 2.4/um M = 5.9
	eSS	C	57 10	LmV B 19 3.3/um 6.0
	LmH	B	17 35.5	
	LmV	B	35.5	
12.	LmH	B	20 40.5	West New Guinea Region 4.28 S 134.25 E
	LmV	B	48.6	H = 19 36 45.8 h = 43 km MB = 5.2 (NEIC)
				D = 113.6
				LmH B 17s 0.9/um M = 5.4
				LmV B 20 0.6/um 5.2
12.	eP	A	22 33 02	Hindu Kush Region 35.61 N 69.70 E
				H = 22 25 03.8 h = 96 km MB = 5.1
				D = 43.84 Az = 308.7 (NEIC)
				PV A 1.2s 20.3nm M = 4.8
13.	e	A	01 22 42	
13.	eP	A	01 42 09.5	North of Ascension Island
	eS	B	50 00	0.95 S 13.20 W
	LmH	B	02 07.5	H = 01 32 36.0 h = normal MB=5.3 MS=4.3
	LmV	B	07.6	D = 55.59 Az = 18.9 (NEIC)
				PV A 1.6s 49.5nm M = 5.3
				LmH B 17 1.1/um 5.0
				LmV B 17 1.4/um 5.2

Day	Phase		h m s	Remarks
13.	eP	A	11 28 36	**Arabian Sea** 14.75 N 55.67 E H = 11 19 36.5 h = 11 km MB = 5.0 D = 50.37 Az = 324.9 (NEIC) PV A 2.4s 82.9nm M = 5.3
13.	eP	A	13 08 51	**Mindoro, Philippine Islands** 13.63 N 120.75 E H = 12 55 42.6 h = 1 km MB = 5.3 D = 91.28 Az = 323.0 (NEIC)
13.	e LmH LmV	A B B	13 31 26 14 18.5 31.4	**Easter Island Region** 23.57 S 111.95 W H = 13 12 11.4 h = normal MB = 5.4 D = 129.02 Az = 43.1 (NEIC) LmH B 22s 0.6/um M = 5.3 LmV B 18 0.6/um 5.4
13.	LmV LmH	B B	13 53.6 54.3	**Mindoro, Philippine Islands** 13.63 N 120.75 E H = 12 55 42.6 h = 1 km MB = 5.3 (NEIC) D = 91.3 LmH B 17.5s 0.5/um M = 5.0 LmV B 16 0.4/um 5.0
13.	ePKIKP LmH LmV	A B B	16 51 01 17 37.7 51.3	**Easter Island Region** 23.70 S 111.98 W H = 16 31 52.9 h = normal MB = 5.3 D = 129.14 Az = 43.1 (NEIC) PKIKPV A traces LmH B 23s 0.5/um M = 5.1 LmV B 18 0.4/um 5.1
14.	+iP	AB	02 30 56.5	**Kurile Islands** 44.06 N 148.22 E H = 02 19 01.6 h = 64 km MB = 5.5 D = 78.41 Az = 333.5 (NEIC) PV A 1.6s 157.0nm M = 5.7
14.	ePKP LmH	AB B	17 30 43 18 49.4	**Fiji Islands Region** 16.65 S 175.87 E H = 17 11 13.8 h = 54 km MB = 5.8

Day	Phase		h m s	Remarks
cont. 14.	LmV	B	18 50.1	D = 143.80 Az = 343.0 (NEIC) PKPV A 1.7s 109.0nm LmH B 17 0.4/um LmV B 17 0.4/um
14.	ePKP	A	20 33 47.5	Tonga 16.96 S 174.61 W H = 20 14 25.5 h = 169 km MB = 4.4 D = 146.02 Az = 353 (ISC) PKPV A 1.6s 13.7nm
14.	ePKIKP ePKHKP ePKP2	A AB A	21 34 36.5 34 43 34 51	South of Fiji Islands 22.00 S 179.12 W H = 21 15 47.9 h = 501 km MB = 5.4 D = 150.19 Az = 346.2 (NEIC) PKIKPV A 1.9s 30.3nm PKHKPV A 2.0 171.0nm PKP2V A 1.1 36.3nm
15.	eP epP	A A	06 26 25 26 38.5	Near East Coast of Honshu, Japan 37.78 N 141.85 E H = 06 14 10.7 h = 50 km MB = 4.8 D = 81.69 Az = 330.6 (NEIC) h = 49 km
15.	eP	A	18 48 31	Afghanistan-USSR Border Region 37.26 N 71.74 E H = 18 40 32.9 h = 126 km MB = 4.9 D = 44.12 Az = 307.5 (NEIC)
15.	e	A	20 15 00.5	
15.	iPn iPg iSn iSg	A A A A	22 51 50.5 52 07 52 33 52 51	Austria 47.17 N 12.91 E H = 22 50 54.7 h = 10 km D = 3.59 Az = 346.7 (NEIC)
16.	ePKHKP	A	08 15 45	South of Fiji Islands 21.97 S 179.39 W H = 07 57 00.6 h = 606 km MB = 4.1 D = 150.10 Az = 345.9 (NEIC) PKHKPV A 1.8s 33.8nm

Day	Phase		h m s	Remarks
16.	ePKHKP	A	19 26 14	South of Fiji Islands 22.00 S 179.45 W H = 19 07 29.4 h = 607 km MB = 5.0 D = 150.12 Az = 345.8 (NEIC)
17.	ePKP	A	02 03 47	Tonga Islands 18.15 S 174.95 W H = 01 44 29.0 h = 215 km MB = 5.0 D = 147.16 Az = 352.3 (NEIC) PKPV A 1.2s 12.2nm
17.	eP	A	06 29 45	Kurile Islands 44.58 N 149.16 E H = 06 17 46.4 h = normal MB = 5.1 D = 78.25 Az = 334.0 (NEIC) PV A 1.0s 21.7nm M = 5.1
17.	e	A	07 31 04.5	Czechoslovakia 50.78 N 14.53 E Explosion 8.7 t H = 07 30.1 (PRU) D = 1.68 Az = 267 (ISC)
17.	+eiP	A	09 46 52	Southern Sinkiang Province, China 41.03 N 82.20 E
	LmV	B	10 08.2	H = 09 38 10.0 h = normal MB = 5.5
	LmH	B	09.6	D = 48.36 Az = 306.5 (NEIC) LmH B 11.5s 1.0/um M = 5.1 LmV B 15 0.8/um 4.9 SKM-III out of operation
17.	eP	A	15 56 50	South of Honshu, Japan 33.14 N 140.67 E
	epP	A	57 05.5	H = 15 44 19.5 h = 62 km MB = 5.5
	ePP	A	16 00 05	D = 85.24 Az = 330.3 (NEIC) h = 55 km
	e	A	00 09	PV A 2.0s 112.0nm M = 5.6
17.	+eiP	A	16 11 47	Colorado 39.79 N 108.37 W H = 16 00 00.0 h = 0 km MB=5.4 MS=4.1 D = 75.73 Az = 34.7 (NEIC) Nuclear explosion of 90 kt (AEC) PV A 0.8s 17.2nm M = 5.2

Day	Phase		h m s	Remarks
17.	+iP	A	16 18 38.5	Iran 35.50 N 57.76 E
	LmV	B	40.6	H = 16 11 37.1 h = 45 km MB = 4.9
	LmH	B	42.3	D = 36.32 Az = 309.2 (NEIC)
				PV A 1.2s 48.0nm M = 5.3
17.	e(pP)	A	19 34 46	Southern Sumatra 2.35 S 100.81 E
				H = 19 21 33.9 h = 80 km MB = 5.1
				D = 91.28 Az = 320.4 (NEIC)
				(h = 51 km)
				(pP)V A 1.6s 16.5nm
18.	ePKHKP	A	09 34 03.5	Fiji Islands Region 20.68 S 178.59 W
	ePKP2	A	34 09.5	H = 09 15 20.2 h = 595 km MB = 4.7
				D = 149.03 Az = 347.3 (NEIC)
18.	+eiP	AB	10 48 48	Kurile Islands 44.59 N 149.33 E
	LmH	B	11 22.3	H = 10 36 50.3 h = normal MB=5.4 MS=5.2
	LmV	B	27.2	D = 78.29 Az = 334.1 (NEIC)
				LmH B 19.5s 1.0/um M = 5.2
				LmV B 18 0.9/um 5.2
				SKM-III out of operation
18.	ePKIKP	A	13 39 08	New Britain Region 6.23 S 151.82 E
	ePP	B	41 00	H = 13 20 10.2 h = 34 km MB=5.8 MS=5.1
	eSP	C	50 55	D = 124.71 Az = 330.3 (NEIC)
	eSPP	C	52 18	LmH B 20s 0.6/um M = 5.3
	eSS	C	58.0	LmV B 18 0.5/um 5.2
	LmH	B	14 34.7	
	LmV	B	35.4	
19.	+eP	AB	00 42 59.5	North Atlantic Ocean 57.55 N 32.99 W
	eS	B	47 35	H = 00 37 22.5 h = normal MB=4.9 MS=4.7
	LmH	B	53.8	D = 26.64 Az = 85.6 (NEIC)
	LmV	B	54.1	PV A 1.6s 46.7nm M = 4.9
				LmH B 15 3.5/um 5.0
				LmV B 14.5 3.9/um 5.2

Day	Phase		h m s	Remarks
19.	eP	A	00 49 48.5	North Atlantic Ocean 57.55 N 33.14 W
				H = 00 44 10.81 h = normal MB = 4.4
				D = 26.72 Az = 85.5 (NEIC)
				PV A 1.8s 20.3nm M = 4.4
19.	eP	A	11 14 56	Afghanistan-USSR Border Region
				36.38 N 71.12 E
				H = 11 06 54.0 h = 110 km MB = 4.5
				D = 44.26 Az = 308.2 (NEIC)
19.	e(P)	A	19 38 02	Central Italy 42.44 N 13.24 E
				H = 19 35 45 h = 5 km
				D = 8.28 Az = 353 (ISC)
19.	eP	A	21 55 58	Eastern Caucasus 42.78 N 45.88 E
	LmH	B	22 06.3	H = 21 50 38.4 h = normal MB = 4.4
	LmV	B	07.0	D = 24.60 Az = 300.6 (NEIC)
20.	ePKP	A	00 55 10	Fiji Region 15.3 S 179.6 W
				H = 00 35 38 h = 0 km MB = 4.6
				D = 143.56 Az = 348
20.	eP	A	03 11 39	Cuba Region 19.80 N 75.40 W
				H = 03 00 09.2 h = normal MB = 4.6
				D = 73.09 Az = 41.6 (NEIC)
20.	LmV	B	14 30.8	Costa Rica 7.98 N 83.02 W
	LmH	B	31.3	H = 13 35 06.8 h = normal MB = 4.4 (NEIC)
				D = 86.8
				LmH B 19s 0.4 μm M = 4.9
				LmV B 21.5 0.7 μm 5.1
20.	+iP	A	18 29 06.6	Southern Alaska 60.97 N 152.44 W
	e	A	29 42	H = 18 18 18.0 h = 118 km MB = 4.9
				D = 68.00 Az = 10.9 (NEIC)
				PV A 1.0s 39.4nm M = 5.3

Day	Phase		h m s	Remarks
21.	ePKHKP	A	05 18 25.5	Fiji Islands Region 20.21 S 177.98 W H = 04 59 38.6 h = 550 km MB = 5.1 D = 148.69 Az = 348.2 (NEIC)
21.	eP	A	09 28 21.5	Kurile Islands 47.51 N 152.37 E H = 09 16 45.1 h = 137 km MB = 4.9 D = 76.55 Az = 335.5 (NEIC)
21.	eP LmH LmV	A B B	20 30 25 21 13.0 13.1	Ryukyu Islands 27.45 N 129.33 E H = 20 17 50.7 h = normal MB = 4.6 D = 84.81 Az = 325.5 (NEIC) PV A traces LmH B 17s 0.3/um M = 4.7 LmV B 16 0.3/um 4.8
21.	ePKP	A	23 27 53	Samoa Islands Region 16.10 S 172.08 W H = 23 08 17.6 h = normal MB = 5.0 D = 145.41 Az = 355.9 (NEIC) PKPV A 1.4s 14.0nm
22.	LmH LmV	B B	01 17.8 19.7	Iran 35.00 N 56.99 E H = 00 50 14.0 h = 37 km MB = 4.0 (NEIC) D = 36.2 LmH B 13s 0.3/um M = 4.2 LmV B 12 0.2/um 4.2
22.	LmH LmV	B B	10 49.6 50.3	LmH B 13s 0.3/um LmV B 12 0.2/um
22.	eP e LmV LmH	A A B B	15 30 00 30 06 36.9 37.1	Southern Greece 36.8 N 22.3 E H = 15 26 14 (BCIS) D = 15.85 LmH B traces LmV B 12s 0.3/um M = 3.9
22.	ePKHKP eX eY	A AB A	17 25 15 25 19.5 25 25	Tonga Islands 20.85 S 174.02 W H = 17 05 25.7 h = normal MB=5.1 MS=5.2 D = 149.94 Az = 352.8 (NEIC)

Day	Phase		h m s	Remarks
cont. 22.	LmH	B	18 43.7	PKHKPV A 1.7s 45.5nm
	LmV	B	45.0	XV A 1.8 101.3nm
				YV A 1.6 90.6nm
				LmH B 17 0.3 μm M = 5.1
				LmV B 16 0.2 μm 5.1
22.	e	A	22 24 11	<u>East New Guinea Region</u>
	e	A	25 42.5	10.01 S 150.32 E
	LmH	B	23 24.7	H = 22 04 58.4 h = 13 km MB=5.5 MS=5.0
	LmV	B	24.8	D = 127.18 Az = 328.2 (NEIC)
				LmH B 18s 0.3 μm M = 5.0
				LmV B 18 0.3 μm 4.9
23.	e	A	03 05 55	
23.	eP	A	10 27 57	<u>Ascension Island Region</u>
	e	A	28 10	4.97 S 11.64 W
	eS	B	36 12	H = 10 17 59.8 h = normal MB=4.8 MS=4.2
	eSS	B	39 50	D = 58.91 Az = 17.1 (NEIC)
	LmV	B	54.6	PV A 1.3s 17.5nm M = 5.0
	LmH	B	55.1	LmH B 16.5 0.5 μm 4.8
				LmV E 17.5 0.8 μm 4.9
23.	e(Pn)	A	22 08 24.5	
	e	A	08 57.5	
	e(Sg)	A	09 16	
24.	eP	A	02 05 28	<u>Philippine Islands Region</u>
	LmV	B	49.0	20.30 N 121.29 E
	LmH	B	50.0	H = 01 52 47.3 h = normal MB = 5.0
				D = 86.29 Az = 323.1 (NEIC)
				PV A 1.5s 15.1nm M = 5.0
				LmV B 15 0.35 μm 4.9
24.	eP	A	02 35 04.5	<u>Mid-Indian Rise</u> 9.17 S 67.21 E
				H = 02 23 16.2 h = normal MB = 4.9
				D = 76.52 Az = 327.3 (NEIC)

Day	Phase		h m s	Remarks
24.	ePKP	A	04 05 46.5	Tonga 15.5 S 174.6 W
				H = 03 46 04.6 h = 0 km
				D = 144.61 Az = 353 (ISC)
24.	(+)eP	AB	18 59 07.5	Andreanof Islands, Aleutian Is.
	epP	A	59 20	51.63 N 173.44 W
	eS	B	19 09 00	H = 18 47 11.7 h = 43 km MB=5.4 MS=5.1
	eSS	B	14 12	D = 78.01 Az = 356.7 (NEIC)
	LmH	B	42.8	h = 46 km
	LmV	B	43.0	PV A 1.0s 41.3nm M = 5.4
				LmH B 19 1.4 μm 5.3
				LmV B 18.5 1.1 μm 5.2
24.	eP	A	19 04 54.5	Andreanof Islands 51.7 N 173.21 W
				H = 18 52 58 h = 26 km MB = 4.4
				D = 77.94 Az = 357 (ISC)
24.	eP	A	19 46 47.5	Guatemala 14.71 N 91.23 W
	epP	A	47 13.5	H = 19 34 13.5 h = 100 km MB = 5.2
				D = 86.70 Az = 38.4 (NEIC)
				h = 100 km
24.	eP	A	21 06 12	Kurile Islands 45.46 N 150.68 E
				H = 20 54 15.8 h = normal MB = 4.8
				D = 77.92 Az = 334.7 (NEIC)
25.	eP	A	08 48 33	West Pakistan 25.54 N 66.50 E
	LmH	B	09 15.0	H = 08 39 53.9 h = 57 km MB = 4.9
	LmV	B	17.1	D = 48.59 Az = 316.0 (NEIC)
				PV A 1.2s 22.4nm M = 5.1
				LmH B 16.5 0.9 μm
				LmV B 13 1.2 μm
25.	eP	A	13 29 12.5	South of Alaska 53.20 N 161.28 W
				H = 13 17 26.1 h = 39 km MB = 4.9
				D = 76.35 Az = 4.6 (NEIC)
				PV A 1.3s 21.8nm M = 5.0

Day	Phase		h m s	Remarks
25.	eP	A	17 08 09	Kurile Islands 45.31 N 151.0 E
				H = 16 56 13 h = 39 km MB = 4.5
				D = 78.16 Az = 335 (ISC)
25.	e(PKHKP)	A	18 06 35.5	South of Fiji Islands
				23.11 S 176.07 W
				H = 17 46 39.5 h = normal MB = 4.9
				D = 151.86 Az = 349.6 (NEIC)
				(PKHKP)V A 1.6s 16.5nm
25.	ePKP	A	21 28 14.5	Fiji Region 19.75 S 177.5 W
	e	A	28 19.5	H = 21 08 31.0 h = 0 km MB = 4.8
				D = 148.34 Az = 349 (ISC)
26.	eP	A	02 27 51	Near East Coast of Honshu, Japan
	epP	A	28 05	37.17 N 141.40 E
	LmH	B	03 07.4	H = 02 15 35.4 h = 56 km MB = 5.0
	LmV	B	09.0	D = 82.04 Az = 330.4 (NEIC)
				h = 50 km
				PV A 1.7s 36.4nm M = 5.1
				LmH B 16 0.2/um
				LmV B 18 0.3/um
26.	eP	A	03 18 46	Andreanof Islands, Aleutian Is.
				51.73 N 175.42 W
				H = 03 06 53.5 h = 59 km MB = 4.6
				D = 77.83 Az = 355.4 (NEIC)
				traces
26.	eP1	A	12 31 29.5	Andreanof Islands, Aleutian Is.
	iP2	AB	31 30.8	51.37 N 179.74 W
	e	B	36 16	H = 12 19 34.4 h = 39 km MB=5.8 MS=5.7
	eS	B	41 16	D = 77.90 Az = 352.6 (NEIC)
	ePPS	B	42 20	P1V A 0.9s 25.3nm M = 5.2
	eSS	C	46 40	P2V A 1.3 111.0nm 5.7
	LmH	B	13 10.1	P2V B 12 1.1/um 5.7
	LmV	B	15.2	LmH B 17.5 5.2/um 5.9
				LmV B 17 4.4/um 5.9

Day	Phase		h m s	Remarks
26.	eP	A	13 24 03.5	Andreanof Islands, Aleutian Is.
				51.28 N 179.68 W
				H = 13 12 10.1 h = 56 km MB = 4.9
				D = 78.00 Az = 352.7 (NEIC)
				PV A 1.2s 16.2nm M = 4.9
27.	ePKIKP	A	06 57 10	Fiji Islands Region 21.33 S 177.93 W
	eiPKHKP	A	57 15.5	H = 06 38 13.4 h = 422 km MB = 5.0
	ePKP2	A	57 22.5	D = 149.79 Az = 347.9 (NEIC)
	epPKHKP	A	58 56	h = 421 km
				PKIKPV A traces
				PKHKPV A 1.2s 34.5nm
				PKP2V A 1.1 18.1nm
27.	eP	A	21 28 26.5	South Indian Ocean 2.54 S 65.15 E
				H = 21 17 17.1 h = normal MB = 4.8
				D = 69.87 Az = 327.0 (NEIC)
				PV A 1.8s 20.3nm M = 4.9
28.	eP	A	20 39 34	Mascarene Islands Region 18.13 S 65.29 E
	e	B	39 52	H = 20 27 11.2 h = normal MB=5.0 MS=5.9
	Pm	A	40 09	D = 83.10 Az = 328.9 (NEIC)
	eS	B	49 52	PV A 2.0s 21.4nm M = 4.9
	eSS	C	55 00	PmV A 1.7 88.0nm 5.7
	LmV	B	18.3	SH B 18 2.1/um 6.0
	LmH	B	22.3	LmH B 20 1.9/um 5.5
				LmV B 18 1.3/um 5.4
29.	eP1	A	01 58 35	Rat Islands, Aleutian Is.
	eP2	AB	58 38.5	51.69 N 176.24 E
	e	A	02 02 52	H = 01 46 44.9 h = 46 km MB=5.2 MS=5.7
	eS	B	08 28	D = 77.21 Az = 350.0 (NEIC)
	eSS	C	13 36	P1V A 1.3s 17.5nm M = 4.9
	LmH	B	35.1	P2V A 2.0 136.8nm 5.9
	LmV	B	42.6	P2V B 10 1.0/um 5.7
				SH B 16 1.5/um 5.8
				LmH B 17.5 3.5/um 5.7
				LmV B 15.5 3.2/um 5.8

Day	Phase		h m s	Remarks
29.	eP	A	04 57 02.5	<u>Greenland Sea</u> 73.71 N 9.55 E
				H = 04 51 57.9 h = normal MB = 4.8
				D = 23.17 Az = 176.7 (NEIC)
				PV A 0.7s 23.0nm M = 4.8
29.	ePKP	A	06 01 41.5	<u>Loyalty Islands</u> 20.68 S 168.51 E
	e	A	01 53	H = 05 42 06.9 h = normal MB = 4.7
				D = 144.98 Az = 334.2 (NEIC)
				PKPV A 1.2s 16.3nm
29.	+iP	AB	06 26 06.0	<u>Unimak Island Region</u> 54.01 N 163.76 W
	eS	B	35 46	H = 06 14 22.3 h = 30 km MB=6.0 MS=5.5
	LmV	B	07 08.8	D = 75.64 Az = 3.0 (NEIC)
	LmH	B	08.9	PV A 1.4s 516.0nm M = 6.4
				PH A 1.8 388.0nm 6.4
				PV B 2.0 0.8/um 6.4
				LmH B 17.5 1.3/um 5.3
				LmV B 17 1.2/um 5.3
29.	e	A	11 59 27	<u>Mascarene Islands Region</u>
				17.45 S 66.50 E
				H = 11 46 54.6 h = normal MB = 4.8
				D = 83.12 Az = 328.4 (NEIC)
				traces
29.	eP	A	23 18 33.5	<u>Komandorsky Islands Region</u>
				55.01 N 163.98 E
				H = 23 07 09.4 h = normal MB = 4.8
				D = 72.21 Az = 341.9 (NEIC)
30.	eP	A	04 50 59	<u>Ecuador</u> 2.26 S 78.52 W
	epP	B	51 26	H = 04 38 01.8 h = 111 km MB = 5.7
	e	A	51 28.5	D = 91.82 Az = 39.6 (NEIC)
	ePP	B	54 44	h = 106 km
	eiSKS	B	05 01 25	PV A 1.3s 17.5nm M = 5.2
	eS	B	01 53	pPV B 8 0.8/um
	e(sS)	B	02 38	PPV B 8 0.6/um 6.0
	eSS	C	08 10	LmH B 19 0.8/um

Day	Phase		h m s	Remarks
cont. 30.	LmV	B	05 23.0	LmV B 20s 1.0/um
	LmH	B	23.1	
30.	ePKP	A	11 59 01.5	Fiji Islands Region 17.61 S 179.05 W
				H = 11 40 21.8 h = 553 km MB = 4.8
				D = 145.95 Az = 347.9 (NEIC)
30.	LmH	B	17 36.7	Northern Easter I. Cordillera
	LmV	B	39.1	4.08 S 104.11 W
				H = 16 32 22.1 h = normal MB=5.0 MS=5.1
				D = 109.4
				LmH B 18s 0.4/um M = 5.0
				LmV B 18 0.3/um 4.9
30.	LmV	B	18 26.2	Gulf of California 26.31 N 110.67 W
	LmH	B	27.3	H = 17 33 51.0 h = normal MB = 4.8
				D = 88.0
				LmV B 16s 0.35/um M = 4.9
31.	eP	A	05 51 55	Near Coast of Guatemala 13.94 N 90.88 W
	e	AB	52 12.5	H = 05 39 18.8 h = 99 km MB = 5.4
	e	A	52 25	D = 87.08 Az = 38.5 (NEIC)
	eSKS	B	06 02 17	PV A 1.7s 42.4nm M = 5.2
	LmV	B	33.7	LmH B 18 0.8/um
	LmH	B	33.8	LmV B 18 0.9/um
31.	e	A	14 47 56	
31.	eP	A	17 48 50.5	Rat Islands, Aleutian Is.
	LmH	B	18 28.0	51.34 N 176.30 E
	LmV	B	32.0	H = 17 36 52.9 h = 10 km MB=4.8 MS=5.0
				D = 77.56 Az = 350.1 (NEIC)
				PV A 1.3s 13.1nm M = 4.9
				LmH B 16 0.4/um 4.8
				LmV B 15 0.2/um 4.7

Day	Phase		h m s	Remarks
31.	eP	A	19 58 16	<u>Southern Iran</u> 28.23 N 56.17 E
				H = 19 50 36.9 h = normal MB = 4.7
				D = 40.26 Az = 316.3 (NEIC)
				PV A 1.4s 7.0nm M = 4.2
31.	eP1	A	23 50 45.5	<u>Burma-India Border Region</u>
	eiP2	AB	51 02.5	24.28 N 93.55 E
	ePP	B	53 28	H = 23 39 56.7 h = 30 km MB=5.9 MS=5.7
	ePPP	B	55 14	D = 66.62 Az = 316.6 (NEIC)
	eS1	C	59 32	P1V A 1.0s 78.7nm M = 5.8
	eS2	C	59 52	P2V A 1.9 583.0nm 6.4
	eSS	C	24 03 54	P2H A 2.0 345.0nm 6.4
	e	C	04 30	P2V B 6 1.7 µm 6.3
	eSSS	C	06 45	LmH B 22.5 8.0 µm 5.9
	LmV	B	18.0	LmV B 24 4.5 µm 5.7
	LmH	B	18.9	
	eP'P'	A	19 14	
	e	A	19 18	

Day	Phase		h m s	Remarks
1.	ePKIKP	A	07 41 54	Southeast Indian Rise 47.80 S 99.66 E
	ePP	A	43 39.5	H = 07 22 57.1 h = normal MB=5.8 MS=6.3
	eSS	C	08 00 20	D = 123.70 Az = 310.1 (NEIS)
	LmH	B	36.8	PKIKPV A 1.2s 8.1nm
	LmV	B	40.4	PPV A 2.4 81.3nm M = 5.8
				LmH B 18.5 3.3/um 6.0
				LmV B 18.5 3.5/um 6.0
1.	e(PKP)	A	08 47 53	South of Australia 50.07 S 114.28 E
				H = 08 28 33.6 h = normal MB = 5.3
				D = 132.82 Az = 302.1 (NEIS)
1.	e	A	12 54 53	Southern Sinkiang Prov., China
	LmH	B	13 17.7	41.13 N 82.09 E
	LmV	B	19.2	H = 12 46 08.8 h = normal MB = 4.9
				D = 48.24 Az = 306.4 (NEIS)
				LmH B 16s 0.7/um M = 4.7
				LmV B 15 0.5/um 4.7
1.	eP	A	13 14 10	Burma-China Border Region
				25.20 N 98.28 E
				H = 13 03 01.8 h = normal MB = 5.1
				D = 69.08 Az = 317.1 (NEIS)
				PV A 1.4s 14.0nm M = 4.8
1.	ePKHKP	A	20 19 10.5	Fiji Islands Region 20.84 S 178.61 W
	ePKP2	A	19 16.5	H = 20 00 28.3 h = 613 km MB = 4.6
				D = 149.18 Az = 347.3 (NEIS)
2.	LmV	B	00 23.6	LmH B 19s 0.4/um
	LmH	B	25.5	LmV B 19 0.4/um
2.	e	A	02 27 55.5	
2.	e(P)	A	20 18 48	Dominican Republic Region
	e	A	18 56	19.58 N 70.61 W
	eS	B	27 52	H = 20 07 30.5 h = 40 km MB=5.2 MS=5.3
	eSS	B	32 32	D = 70.24 Az = 42.1 (NEIS)

Day	Phase		h m s	Remarks
cont. 2.	LmH	B	20 49.5	PV A 2.0s 34.2nm M = 5.0
	LmV	B	49.5	LmH B 17.5 1.8/um 5.4
				LmV B 17.5 1.9/um 5.4
2.	eP1	AB	24 05 38.5	Northern Sinkiang Prov., China
	-iP2	AB	05 46	44.10 N 83.55 E
	ePP	B	07 35	H = 23 57 04.2 h = 26 km MB=5.8 MS=5.6
	eS	B	12 30	D = 47.39 Az = 304.7 (NEIS)
	eScS	C	15 30	P1V A 1.1s 101.0nm M = 5.8
	eSS	C	15 48	P2V A 1.6 522.0nm 6.3
	eSa	B	16 18	P2V B 5 1.3/um 6.8
	eLg1	B	22 00	LmH B 14.5 12.2/um 6.0
	LmH	B	26.4	LmV B 13.5 17.1/um 6.3
	LmV	B	26.4	Sa : velocity 4.53 km/s
				Lg1: velocity 3.52 km/s
3.	LmV	B	10 04.8	LmH B 18s 0.3/um
	LmH	B	05.9	LmV B 18 0.3/um
4.	eP	A	02 55 54	South of Honshu, Japan 33.26 N 140.40 E
	epP	A	56 09	H = 02 43 24.3 h = 62 km MB = 4.4
				D = 85.02 Az = 330.1 (NEIS)
				h = 53 km
				PV traces
4.	ePKIKP	A	09 42 26.5	South of Fiji Islands 23.45 S 179.90 W
	ePKHKP	A	42 33	H = 09 23 40.1 h = 554 km MB = 5.0
	ePKP2	A	42 43.5	D = 151.41 Az = 345.6 (NEIS)
				PKIKPV traces
4.	LmH	B	16 17.8	Near East Coast of Eastern Russia
	LmV	B	20.5	46.64 N 140.85 E
				H = 15 34 39.8 h = normal MB = 4.6 (NEIS)
				D = 73.7
5.	eP	A	01 28 23	West Pakistan 25.10 N 68.07 E
	e	A	28 26.5	H = 01 19 30.3 h = normal MB = 4.8
				D = 49.89 Az = 316.1 (NEIS)
				PV A 0.8s 7.7nm M = 4.8

Day	Phase		h m s	Remarks
5.	eP	A	03 10 00.5	Unimak Island Region 54.04 N 164.04 W
				H = 02 58 15.3 h = 17 km MB = 4.8
				D = 75.63 Az = 2.9 (NEIS)
5.	ePKIKP	B	03 31 50	New Hebrides Islands 17.19 S 167.81 E
	e	A	31 58	H = 03 12 25.8 h = 24 km MB=5.6 MS=6.1
	e	A	32 05	D = 141.57 Az = 335.6 (NEIS)
	ePP	B	35 04	PKIKPV B 8s 2.1/um
	eSS	B	53 32	PPV B 9 1.9/um
	eSSS	C	58 50	LmH B 19 2.9/um M = 6.0
	LmV	B	41.6	LmV B 18 3.3/um 6.1
	LmH	B	42.9	
5.	ePKIKP	A	07 55 07.5	New Hebrides Islands 17.36 S 167.70 E
				H = 07 35 35.3 h = 15 km MB = 4.7
				D = 141.67 Az = 335.4 (NEIS)
				PKIKPV traces
5.	ePKIKP	A	09 24 05	New Hebrides Islands 17.34 S 167.79 E
				H = 09 04 32.9 h = 21 km MB = 5.0
				D = 141.69 Az = 335.5 (NEIS)
				PKIKPV A 1.4s 9.3nm
5.	e	A	13 49 06.5	Northern Italy 44.67 N 9.52 E
	ePg	A	49 24.5	H = 13 47 24.1 h = normal MB = 4.1 (NEIS)
	eSn	A	50 00	D = 6.15
	eiSg	AB	50 46	LmH B 10s 4.8/um M = 4.3
	LmH	B	51.4	LmV B 6.5 0.7/um
	LmV	B	52.1	
5.	epP	A	23 04 51	Near East Coast of Honshu, Japan
	LmH	B	24 03.6	38.39 N 142.24 E
	LmV	B	03.6	H = 22 52 26.7 h = 44 km MB = 4.7
				D = 81.31 Az = 330.7 (NEIS)
				h = 40 km
				LmH B 13s 0.2/um M = 4.7
				LmV B 13 0.2/um 4.7

Day	Phase		h m s	Remarks
6.	+iP	AB	13 12 18.0	Southern Nevada 37.25 N 116.35 W
	e	A	12 36	H = 13 00 00.1 h = 0 km MB = 6.1
	ePP	A	15 22.5	D = 81.25 Az = 30.5 (NEIS)
	LmH	B	51.0	PV A 1.4s 209.0nm M = 6.0
	LmV	B	51.0	PH A 1.8 105.5nm 6.0
				PV B 2.5 0.6 µm 6.1
				PPV A 1.6 93.4nm 5.8
				LmH B 16 0.6 µm 5.0
				LmV B 15 0.8 µm 5.2
6.	iP	A	15 59 04.7	Hindu Kush Region 36.33 N 70.64 E
				H = 15 51 17.1 h = 220 km MB = 5.1
				D = 43.99 Az = 308.2 (NEIS)
6.	ePn	A	20 39 02	Northern Italy 45.55 N 12.01 E
	eiPg	A	39 25	H = 20 37 45.6 h = 30 km MB = 4.0
	eSn	A	39 58	D = 5.11 Az = 357.2 (NEIS)
	eiSg	A	40 28.5	
6.	ePn	A	21 13 40	Yugoslavia 42.38 N 18.65 E
	eSg	A	16 30	H = 21 11 20.5 h = 16 km MB = 4.6
	e	A	16 33	D = 9.58 Az = 332.1 (NEIS)
	LmH	B	16.9	LmH B 9s 1.0 µm M = 4.0
	LmV	B	17.5	LmV B 8 1.0 µm
6.	LmV	C	22 55.5	Probably New Hebrides Islands
				17.45 S 167.61 E
				H = 21 26 40.5 h = 5 km (NEIS)
				D = 141.6
				LmV C 18s 0.3 µm M = 5.1
7.	ePKP2	A	03 04 06	Macquarie Islands Region
	LmH	C	04 33.0	53.93 S 159.40 E
	LmV	C	33.0	H = 02 43 31.4 h = normal MB = 5.8 (NEIS)
				D = 160.12
				PKP2V A 1.4s 20.9nm
				LmH C 18 0.3 µm M = 5.1
				LmV C 18 0.45 µm 5.3

Day	Phase		h m s	Remarks
7.	eP	A	06 34 36	<u>Taiwan Region</u> 22.36 N 121.11 E
	LmH	B	07 12.8	H = 06 22 01.9 h = normal MB = 4.4
	LmV	B	16.9	D = 84.54 Az = 322.9 (NEIS)
				PV A traces
				LmH B 13s 0.7/um M = 5.2
				LmV B 14 0.7/um 5.2
7.	+eP1	AB	18 45 23.5	<u>Near Coast of Chiapas, Mexico</u>
	eP2	AB	47 29	14.28 N 92.01 W
	epP1	AB	47 42	H = 18 32 42.9 h = 78 km MB = 5.5
	ePP1	B	48 48	D = 87.50 Az = 38.3 (NEIS)
	ePP2	B	51 00	P1V A 2.4s 110.5nm M = 5.6
	e	B	51 15	P1V B 13 1.3/um 5.9
	eS1	C	55 45	P2V A 1.0 27.6nm 5.3
	eS2	C	57 55	P2V B 9 1.4/um 6.1
	LmH	B	19 29.6	pP1V A 2.8 192.0nm
	LmV	B	29.7	P1V B 10 1.8/um
				LmH B 18 18.3/um
				LmV B 18 13.4/um
7.	ePKIKP	A	19 15 05	<u>South of Fiji Islands</u> 22.15 S 176.75 W
	-iPKHKP	AB	15 11.2	H = 18 55 43.4 h = 220 km MB = 5.2
	ePKP2	A	15 18	D = 150.81 Az = 349.0 (NEIS)
	epPKP	A	16 00	PKHKPV A 1.8s 115.0nm
				PKP2V A 1.5 42.7nm
				pPKPV A 1.8 60.8nm
8.	eP	A	00 35 03	<u>Kurile Islands</u> 45.41 N 151.02 E
				H = 00 23 05.4 h = normal MB = 4.8
				D = 78.06 Az = 334.9 (NEIS)
8.	ePKIKP	A	01 20 38.5	<u>New Hebrides Islands</u> 17.47 S 167.74 E
	ePP	C	23 50	H = 01 01 11.4 h = 21 km MB= 5.2
	LmH	B	02 31.5	MS = 5.5 (NEIS)
	LmV	B	31.5	D = 141.79
				PKIKPV A 1.2s 12.2nm
				LmH B 18 0.6/um M = 5.4
				LmV B 18 0.5/um 5.3

Day	Phase		h m s	Remarks
8.	eP	A	10 37 02	Near East Coast of Honshu, Japan
				36.15 N 140.00 E
				H = 10 24 48.1 h = 80 km MB = 4.8
				D = 82.36 Az = 329.8 (NEIS)
				traces
8.	LmH	B	18 29.1	Southern Iran 26.46 N 61.12 E
	LmV	B	33.8	H = 17 56 59.1 h = normal MB = 4.6
				D = 44.57 Az = 316.4 (NEIS)
				LmH B 16s 0.5 μm M = 4.5
				LmV B 14 0.3 μm 4.5
8.	LmH	B	22 22.6	Southern Iran 26.6 N 61.0 E
	LmV	B	24.3	H = 21 49 56.8 h = normal MB = 4.6 (NEIS)
				D = 44.4
				LmH B 16s 0.4 μm M = 4.4
				LmV B 13 0.3 μm 4.5
9.	eiP	A	01 48 47	Near East Coast of Kamchatka
	LmH	B	02 17.1	53.93 N 160.52 E
	LmV	B	19.5	H = 01 37 21.4 h = 33 km MB=5.3 MS=5.0
				D = 72.58 Az = 339.8 (NEIS)
				PV A 1.4s 58.0nm M = 5.4
				LmH B 20.5 1.5 μm 5.2
				LmV B 18 1.7 μm 5.4
9.	eP	A	03 55 13.5	Kurile Islands Region 44.06 N 151.66 E
	e	A	55 17.5	H = 03 43 09.1 h = 34 km MB = 5.0
				D = 79.48 Az = 335.4 (NEIS)
				PV A 1.2s 8.1nm M = 4.6
9.	eP	A	04 27 57	Southern Sinkiang Prov., China
	LmV	B	49.3	41.05 N 82.31 E
	LmH	B	50.6	H = 04 19 14.3 h = normal MB = 5.1
				D = 48.42 Az = 306.6 (NEIS)
				PV A 1.4s 18.6nm M = 4.9
				LmH B 13 0.6 μm 4.8
				LmV B 14 0.4 μm 4.7

Day	Phase		h m s	Remarks
9.	eP	A	08 28 19	Kansu Province, China 39.42 N 95.41 E
	e	A	28 22.5	H = 08 18 32.5 h = normal MB = 5.0
	e	A	28 29	D = 57.25 Az = 311.2 (NEIS)
				PV A 1.5s 20.1nm M = 5.0
9.	ePKIKP	AB	08 40 34	Solomon Islands 10.29 S 161.36 E
	epPKIKP	AB	40 56	H = 08 21 27.3 h = 70 km MB = 6.3
	ePP	AB	43 00	D = 132.70 Az = 334.1 (NEIS)
	epPP	B	43 18	h = 75 km
	eSKP	B	43 55	PKIKPV A 2.8s 193.0nm
	e	B	44 20	pPKIKPV A 2.1 288.0nm
	ePPS	B	54 48	PPV A 3.5 833.0nm M = 6.1
	eSS	B	09 00 24	PPV B 6 1.6 μm 6.2
	e	C	02 24	LmH B 21.5 8.2 μm 6.4
	eSSS	C	05 44	LmV B 21.5 6.8 μm 6.3
	LmH	B	39.0	
	LmV	B	39.8	
9.	eP	A	19 13 50	Dodecanese Islands 36.22 N 28.38 E
				H = 19 09 33.1 h = 58 km MB = 4.1
				D = 18.81 Az = 325.3 (NEIS)
				PV A 1.7s 54.5nm M = 4.5
9.	eP	A	22 09 05.5	Andaman Islands Region 11.58 N 95.07 E
	e	A	09 14	H = 21 57 14.1 h = normal MB=5.0 MS=4.8
				D = 77.00 Az = 319.5 (NEIS)
				PV A 1.2s 16.3nm M = 4.9
9.	+iP	AB	22 57 29.8	Off East Coast of Kamchatka 52.91N 160.08E
	LmH	B	23 31.2	H = 22 45 58.9 h = normal MB = 5.4 (NEIS)
	LmV	B	36.0	D = 73.44
				PV A 1.0s 134.0nm M = 5.9
				LmH B 18.5 0.6 μm 4.9
				LmV B 16 0.6 μm 5.0
9.	e(P)	A	24 10 17	Near East Coast of Kamchatka
				53.14 N 160.01 E
				H = 23 58 47.2 h = normal MB = 4.6
				D = 73.22 Az = 339.6 (NEIS)

Day	Phase		h m s	Remarks
10.	LmH	B	08 52.0	LmV B 18s 0.3/um
	LmV	B	52.0	
10.	eP	A	16 16 53.5	Southern Sinkiang Prov., China
	LmH	B	36.0	39.53 N 74.81 E
	LmV	B	38.9	H = 16 08 42.2 h = normal MB = 5.2
				D = 44.69 Az = 306.1 (NEIS)
				PV A 1.1s 20.2nm M = 4.9
				LmH B 18 0.8/um 4.5
				LmV B 16 0.8/um 4.6
10.	LmH	C	17 15.0	LmH C 46s 0.7/um
	LmV	C	19.0	
11.	eiPn	A	03 17 01.3	Yugoslavia 46.19 N 16.09 E
	ePg	A	17 20	H = 03 15 39.3 h = 11 km MB = 4.5
	eSn	A	17 59	D = 5.36 Az = 327.9 (NEIS)
	eiSg	AB	18 30	PnV A 1.1s 60.5nm M = 5.1
	LmH	B	19.0	LmH B 9 1.4/um 3.7
	LmV	B	19.0	LmV B 9 1.4/um
11.	eiP	AB	08 53 32.5	Off East Coast of Kamchatka
	ei	AB	53 51	53.69 N 161.59 E
	eS	B	09 03 00	H = 08 42 04.0 h = 30 km MB=5.6 MS=6.0
	ei	B	03 16	D = 73.01 Az = 340.5 (NEIS)
	LmV	B	31.5	PV A 1.6s 270.0nm M = 6.0
	LmH	B	33.3	PV B 5 2.9/um 6.8
				LmH B 17 12.0/um 6.2
				LmV B 16.5 9.3/um 6.2
11.	eP	A	20 14 08	Afghanistan-USSR Border Region
				36.75 N 71.30 E
				H = 20 06 07.1 h = 97 km MB = 4.9
				D = 44.16 Az = 307.9 (NEIS)
11.	e	A	20 40 40.5	

Day	Phase		h m s	Remarks
12.	eP	A	00 37 42	Off East Coast of Kamchatka
	e(PcP)	A	37 55	53.69 N 161.64 E
				H = 00 26 14.5 h = normal MB = 4.2
				D = 73.03 Az = 340.6 (NEIS)
12.	eP	A	01 20 59	Off East Coast of Kamchatka
				53.71 N 161.67 E
				H = 01 09 30.7 h = normal MB = 4.2
				D = 73.01 Az = 340.6 (NEIS)
12.	eP	A	01 45 50	Off East Coast of Kamchatka
	e(PcP)	A	46 06.5	53.65 N 161.76 E
				H = 01 34 20.6 h = normal MB = 4.2
				D = 73.09 Az = 340.7 (NEIS)
12.	eP	A	03 48 25	Off East Coast of Kamchatka
	e(PcP)	A	48 38	53.77 N 161.69 E
				H = 03 36 56.4 h = normal MB = 4.5
				D = 72.96 Az = 340.6 (NEIS)
12.	eP	A	06 55 30	Off East Coast of Kamchatka
	e(PcP)	A	55 43.5	53.64 N 161.63 E
	LmH	C	07 28.0	H = 06 44 02.0 h = normal MB=4.8 MS=4.3
	LmV	C	34.4	D = 73.07 Az = 340.6 (NEIS)
				PV A 1.0s 15.8nm M = 5.0
				LmV C 16 0.6/um 5.0
12.	eP	A	11 06 17.5	Crete 34.22 N 26.19 E
	e	A	06 39.5	H = 11 01 52.6 h = 54 km MB = 4.3
				D = 19.57 Az = 331.4 (NEIS)
12.	+eP	AB	14 32 53	Off East Coast of Kamchatka
	e(PcP)	A	33 07	53.63 N 161.59 E
	eS	B	42 36	H = 14 21 24.2 h = normal MB=5.4 MS=5.2
	LmH	B	15 05.8	D = 73.08 Az = 340.5 (NEIS)
	LmV	B	11.7	PV A 2.0s 214.0nm M = 5.8
				PV B 5 0.7/um 5.9
				LmH B 19.5 2.3/um 5.5
				LmV B 16 1.9/um 5.5

Day	Phase		h m s	Remarks
12.	epP	A	15 37 06.5	Hokkaido, Japan Region 41.6 N 139.2 E
	e	A	38 56	H = 15 24 42.1 h = 199 km MB = 4.8 (NEIS)
				D = 77.34
				h = 210 km
12.	ePKP2	A	19 23 54	South of Fiji Islands 25.82 S 177.43 W
	LmH	B	20 42.7	H = 19 03 40.9 h = 45 km MB = 4.7
	LmV	B	43.0	D = 154.26 Az = 346.7 (NEIS)
12.	ePn	A	21 04 02	Austria 47.57 N 15.33 E
	ePg	A	04 13	H = 21 02 56.9 h = 5 km MB = 4.8
	eiSn	A	04 45	D = 3.93 Az = 323.0 (NEIS)
	eiSg	AB	05 01	
12.	eP	A	23 24 27	North Atlantic Ridge 46.57 N 27.47 W
	LmH	B	34.6	H = 23 18 55.5 h = normal MB=4.7 MS=4.6
	LmV	B	35.9	D = 25.95 Az = 66.5 (NEIS)
				PV A 1.8s 20.3nm M = 4.4
13.	eP	A	00 32 25.5	Kurile Islands 46.97 N 150.99 E
	LmH	B	01 10.0	H = 00 20 49.5 h = 142 km MB = 5.5
	LmV	B	11.0	D = 76.64 Az = 334.8 (NEIS)
				PV A 2.0s 205.0nm M = 5.5
13.	eP	A	06 38 57.5	Off East Coast of Honshu, Japan
				39.79 N 143.31 E
				H = 06 26 47.3 h = 32 km MB = 4.6
				D = 80.49 Az = 331.2 (NEIS)
				PV A 1.0s 7.9nm M = 4.7
13.	ePKP	AB	07 01 00	New Hebrides Islands 20.04 S 169.69 E
	LmV	B	08 04.2	H = 06 41 27.3 h = 41 km MB = 5.5 MS=5.0
	LmH	B	04.7	D = 144.89 Az = 335.6 (NEIS)
				PKPV A 2.0s 85.5nm
				LmH B 21 0.5/um M = 5.2
				LmV B 20 0.4/um 5.2

Day	Phase		h m s	Remarks
13.	LmV	B	09 34.0	Revilla Gigedo Islands Region
	LmH	B	34.2	19.69 N 109.50 W
				H = 08 39 26.8 h = normal MS = 5.1 (NEIS)
				D = 93.0
				LmH B 16s 0.9 μm M = 5.4
				LmV B 18 0.9 μm 5.3
13.	+ePKP	A	10 07 58	New Hebrides Islands 19.18 S 169.67 E
				H = 09 48 25.2 h = 24 km MB = 5.7
				D = 144.10 Az = 336.1 (NEIS)
				PKPV A 1.6s 35.7nm
13.	eP	A	13 52 22	Off East Coast of Kamchatka
				53.71 N 161.79 E
				H = 13 40 53.0 h = normal MB = 3.9
				D = 73.03 Az = 340.7 (NEIS)
13.	eP	A	16 12 26	Off East Coast of Kamchatka
				53.5 N 161.2 E
				H = 16 01 01 h = 70 km MB = 4.0
				D = 73.14 Az = 340 (ISC)
14.	ePKIKP	A	03 50 10	Fiji Islands Region 21.09 S 178.78 W
	ePKHKP	A	50 15	H = 03 31 30.9 h = 593 km MB = 5.2
	ePKP2	A	50 22	D = 149.39 Az = 347.0 (NEIS)
	epPKP	A	52 36	
14.	ePg	A	05 56 42	Swabian Jura Region, Federal Rep.of Germany
	eiSg	A	57 16.5	48.7 N 9.0 E
				H = 05 55 51 (BCIS)
				D = 2.59
14.	epP	A	11 18 22	Flores Sea 7.29 S 120.39 E
	ePKP	A	20 05	H = 11 02 46.9 h = 631 km MB = 5.8 (NEIS)
	e	BC	23 40	D = 107.48
	e	BC	28 56	
14.	LmV	B	11 35.1	LmH B 18s 0.5 μm
	LmH	B	38.9	LmV B 14 0.6 μm

Day	Phase		h m s	Remarks
14.	eP	A	12 56 58.5	<u>North of Severnaya Zemlya</u> 84.14 N 113.77 E H = 12 49 16.2 h = normal MB = 4.5 D = 41.13 Az = 288.9 (NEIS) PV A 1.3s 13.1nm M = 4.5
14.	eP	A	18 28 19	<u>Hindu Kush Region</u> 36.40 N 70.80 E H = 18 20 29.3 h = 203 km MB = 5.0 (NEIS) D = 44.05 PV A 1.7s 30.3nm M = 4.5
14.	eP	A	19 57 21.5	<u>North of Severnaya Zemlya</u> 84.01 N 113.08 E H = 19 49 38.5 h = normal MB = 4.7 D = 41.10 Az = 288.3 (NEIS) PV A 1.1s 10.1nm M = 4.5
15.	eP	A	01 18 23	<u>Maine</u> 45.32 N 70.91 W H = 01 09 04.2 h = 12 km MB = 4.8 D = 52.78 Az = 52.4 (NEIS) PV A 1.6s 16.5nm M = 4.7
15.	eP1	AB	11 32 13	<u>Off East Coast of Kamchatka</u>
	+iP2	AB	32 20.0	53.53 N 161.44 E
	eS1	B	41 42	H = 11 20 44.9 h = normal MB = 5.0
	eS2	B	42 00	D = 73.14 Az = 340.5 (NEIS)
	LmH	B	11.3	and
	LmV	B	11.3	<u>Off East Coast of Kamchatka</u> 53.53 N 161.50 E H = 11 20 51.5 h = normal MB = 5.6 MS = 5.9 (NEIS) D = 73.15 P1V A 1.0s 49.2nm M = 5.5 P2V A 2.0 402.0nm 6.1 P2H A 2.0 245.0nm 6.1 P2V B 5 2.6 µm 6.8 S2H B 10 4.0 µm 6.6 LmH B 17 10.2 µm 6.2 LmV B 15.5 12.4 µm 6.3

Day	Phase		h m s	Remarks
15.	eP	A	11 51 10.5	Off East Coast of Kamchatka 53.66 N 161.33 E H = 11 39 42.0 h = normal MB = 4.1 D = 72.99 Az = 340.4 (NEIS)
15.	+iP	A	12 22 56.7	Andreanof Islands, Aleutian Is. 51.30 N 179.39 W H = 12 11 02.3 h = 48 km MB=5.8 MS=4.8 D = 78.00 Az = 352.9 (NEIS) PV A 1.1s 161.5nm M = 5.9
15.	+eP	A	13 28 24	Off East Coast of Kamchatka 53.64 N 161.80 E H = 13 16 56.1 h = 39 km MB=5.1 MS=5.1 D = 73.10 Az = 340.7 (NEIS) PV A 1.2s 30.5nm M = 5.1
15.	+iP	A	13 50 17.8	Andreanof Islands, Aleutian Is. 51.27 N 179.42 W H = 13 38 23.1 h = 50 km MB = 5.4 D = 78.03 Az = 352.8 (NEIS) PV A 1.4s 46.5nm M = 5.3
15.	eP LmH LmV	A B B	19 31 05.5 20 08.2 08.4	Off Coast of Northern California 41.22 N 125.54 W H = 19 18 51.7 h = normal MB=5.2 MS=4.1 D = 81.14 Az = 26.0 (NEIS) PV A 1.0s 11.8nm M = 4.8 LmH B 17.5 0.6/um 5.0 LmV B 16 0.6/um 5.1
15.	eP	A	21 08 02	Off East Coast of Kamchatka 53.65 N 161.81 E H = 20 56 33.9 h = 35 km MB = 4.9 D = 73.10 Az = 340.7 (NEIS) PV A 1.2s 12.2nm M = 4.8

Day	Phase		h m s	Remarks
15.	+iP	AB	21 21 09.3	Off East Coast of Kamchatka
	e	A	21 17	53.56 N 161.55 E
	eS	B	30 52	H = 21 09 41.7 h = 45 km MB=5.5 MS=5.1
	LmH	B	53.9	D = 73.13 Az = 340.5 (NEIS)
	LmV	B	22 00.1	PV A 1.5s 141.0nm M = 5.7
				LmH B 19 2.0/um 5.4
				LmV B 17 2.0/um 5.5
15.	LmH	B	23 12.5	LmH B 19s 1.0/um
	LmV	B	16.6	LmV B 20 0.9/um
15.	ePKIKP	A	23 24 40.5	South of Fiji Islands 25.92 S 177.41 W
	ePKHKP	A	24 49.5	H = 23 04 58.6 h = 94 km MB = 5.6
	ePKP2	A	25 03	D = 154.35 Az = 346.7 (NEIS)
16.	eP	A	01 06 31.5	Hokkaido, Japan Region 43.03 N 145.87 E
				H = 00 54 34.2 h = 54 km MB = 4.8
				D = 78.55 Az = 332.3 (NEIS)
16.	eP	A	01 48 45.5	Hokkaido, Japan Region 42.86 N 144.64 E
				H = 01 36 51.9 h = 70 km MB = 4.9
				D = 78.28 Az = 331.6 (NEIS)
16.	eP	A	07 32 43	Tsinghai Province, China
	LmH	B	58.7	37.71 N 95.64 E
	LmV	B	59.4	H = 07 22 48.1 h = normal MB = 5.4
				D = 58.52 Az = 312.1 (NEIS)
				PV A 1.4s 27.9nm M = 5.2
				LmH B 14 0.7/um 4.9
				LmV B 15 0.9/um 5.1
16.	eP	A	12 22 09	Lake Baikal Region 55.00 N 112.57 E
	e	A	22 23	H = 12 12 32.2 h = normal MB = 4.5
	LmH	B	48.3	D = 55.90 Az = 311.0 (NEIS)
	LmV	B	49.9	PV A 1.3s 13.1nm M = 4.8
				LmH B 18 0.8/um 4.9
				LmV B 14 0.9/um 5.1

Day	Phase		h m s	Remarks
16.	-eP	AB	14 55 44	Off Coast of Oregon 44.98 N 125.77 W
	eS	C	15 06 00	H = 14 43 47.5 h = normal MB=5.6 MS=5.1
	eSS	C	10 50	D = 77.84 Az = 26.2 (NEIS)
	LmH	B	34.0	PV A 1.5s 90.5nm M = 5.6
	LmV	B	34.0	LmH B 14.5 1.9 /um 5.6
				LmV B 14 2.4 /um 5.7
16.	eP	A	16 29 12	Andreanof Islands, Aleutian Is.
				51.43 N 179.30 W
				H = 16 17 18.3 h = 51 km MB = 4.4
				D = 77.89 Az = 352.9 (NEIS)
16.	eP	A	18 46 57	Off East Coast of Kamchatka
	epP	A	47 10	53.62 N 161.49 E
				H = 18 35 29.4 h = 45 km MB = 4.7
				D = 73.06 Az = 340.5 (NEIS)
				h = 50 km
16.	eP	A	20 00 59	Off East Coast of Kamchatka
	epP	A	01 13	53.52 N 161.46 E
				H = 19 49 31.0 h = 46 km MB = 4.9
				D = 73.15 Az = 340.5 (NEIS)
				h = 50 km
16.	eP	A	20 39 37.5	Near East Coast of Honshu, Japan
				35.94 N 140.12 E
				H = 20 27 21.0 h = 76 km MB = 4.6
				D = 82.60 Az = 329.8 (NEIS)
16.	eP	A	20 45 38	Off East Coast of Kamchatka
				53.57 N 161.42 E
				H = 20 34 10.6 h = 47 km MB = 4.8
				D = 73.10 Az = 340.4 (NEIS)
				PV A 1.0s 11.8nm M = 4.8
16.	ePKIKP	A	23 25 01	South Sandwich Islands Region
	LmH	B	24 29.4	58.28 S 25.52 W

Day	Phase		h m s	Remarks
cont. 16.	LmV	B	24 29.4	H = 23 06 28.8 h = 50 km MB = 5.4
				D = 112.74 Az = 24.6 (NEIS)
				LmH B 15.5s 0.7/um
				LmV B 16 0.8/um
17.	+iP1	AB	04 06 59.5	Hokkaido, Japan Region 43.23 N 145.79 E
	Pm	A	07 28	H = 03 55 02.9 h = 48 km MB=6.5 MS=7.7
	eiP2	B	07 30	D = 78.34 Az = 332.2 (NEIS)
	+eS1	BE	16 50	P1V A 2.0s 1472.0nm M = 6.6
	+iS1	BN	16 59	PmV A 1.5 1508.0nm 6.8
	ei(S2)	B	17 30	P2V B 19 127.4/um 7.6
	eSS	B	23 30	LmH B 15 675.0/um 8.1
	eP'P'	A	34 05.5	LmV B 17 670.0/um 8.1
	LmH	B	44.9	possibly multiple earthquake
	LmV	B	48.6	
17.	+iP	A	04 21 13	Kurile Islands 43.00 N 146.41 E
				H = 04 09 12.44 h = normal MB = 5.9
				D = 78.76 Az = 332.6 (NEIS)
				PV A 1.7s 133.3nm M = 5.8
17.	eP	A	04 31 34	Hokkaido Japan Region 42.43 N 145.81 E
				H = 04 19 32.2 h = 35 km MB = 5.4
				D = 79.06 Az = 332.3 (NEIS)
17.	eP	A	05 22(58)	Kurile Islands 43.49 N 146.16 E
	e	A	22 06.5	H = 05 09 58.5 h = 29 km MB = 4.8
				D = 78.24 Az = 332.4 (NEIS)
17.	eP	A	05 24 10	Off Coast of Hokkaido, Japan
	e	A	24 24.5	42.67 N 146.12 E
				H = 05 12 09.3 h = 44 km MB = 5.4
				D = 78.96 Az = 332.4 (NEIS)
				PV A 1.6s 30.2nm M = 5.0
17.	eP	A	06 04 10	Kurile Islands 43.08 N 146.14 E

Day	Phase		h m s	Remarks
cont. 17.	epP	A	06 04 20	H = 05 52 10.9 h = 44 km MB = 5.1 D = 78.60 Az = 332.4 (NEIS) h = 36 km PV A 1.0s 25.6nm M = 5.2
17.	e(P)	A	06 41 30	<u>Kurile Islands</u> 43.38 N 146.18 E H = 06 29 28.4 h = normal MB = 4.4 D = 78.34 Az = 332.4 (NEIS)
17.	-eP	A	07 55 00	<u>Off East Coast of Kamchatka</u>
	e	A	55 10	53.48 N 161.54 E
	e	A	55 27	H = 07 43 31.6 h = 45 km MB = 5.0 D = 73.21 Az = 340.5 (NEIS) PV A 0.9s 19.5nm M = 5.0
17.	eP	A	08 29 48	<u>Kurile Islands</u> 43.01 N 146.42 E H = 08 17 47.7 h = 41 km MB = 5.2 D = 78.76 Az = 332.6 (NEIS) PV A 1.8s 20.3nm M = 5.1
17.	eP	A	09 00 20.5	<u>Off Coast of Hokkaido, Japan</u> 42.91 N 146.46 E H = 08 48 19.4 h = 37 km MB = 5.2 D = 78.86 Az = 332.6 (NEIS)
17.	eP	A	11 11 21	<u>Kurile Islands</u> 42.95 N 146.95 E H = 10 59 18.9 h = 37 km MB = 4.8 D = 78.98 Az = 332.9 (NEIS) traces
17.	+iP	A	12 26 24.5	<u>Hokkaido, Japan Region</u>
	epP	A	26 41	42.92 N 145.57 E
	esP	A	26 48	H = 12 14 25.8 h = 41 km MB = 5.5 D = 78.55 Az = 332.1 (NEIS) h = 64 km PV A 1.3s 35.0nm M = 5.3

Day	Phase		h m s	Remarks
17.	eiP	A	12 36 37	Off Coast of Hokkaido, Japan
	e	A	36 49	42.62 N 146.45 E
	eS	B	46 32	H = 12 24 34.8 h = 36 km MB=5.5 MS=5.0
	LmH	B	13 14.8	D = 79.11 Az = 332.6 (NEIS)
	LmV	B	15.3	PV A 1.4s 32.6nm M = 5.1
				LmH B 17 2.3/um 5.6
				LmV B 18 1.9/um 5.5
17.	eiP1	AB	13 45 24	Hokkaido, Japan Region 43.07 N 145.38 E
	iP2	A	45 30	H = 13 33 28.3 h = 46 km MB=5.6 MS=5.3
	eS	B	55 15	D = 78.35 Az = 332.0 (NEIS)
	LmH	B	14 33.0	P1V A 2.0s 205.0nm M = 5.7
	LmV	B	34.2	LmH B 17 3.0/um 5.7
				LmV B 17 3.6/um 5.8
17.	+eiP	A	13 55 09.5	Kurile Islands 42.95 N 146.66 E
	i	A	55 20	H = 13 43 08.7 h = 40 km MB=5.5 MS=5.2
				D = 78.89 Az = 332.7 (NEIS)
				PV A 1.6s 66.5nm M = 5.4
17.	eP	A	14 36 13	Hokkaido, Japan Region 42.90 N 145.32 E
				H = 14 24 13.8 h = 34 km MB = 4.9
				D = 78.48 Az = 332.0 (NEIS)
17.	eP	A	17 05 14	Rat Islands, Aleutian Is.
				51.75 N 176.37 E
				H = 16 53 24.3 h = 53 km MB = 5.0
				D = 77.17 Az = 350.1 (NEIS)
				PV A 0.7s 11.5nm M = 5.0
17.	+iP	AB	19 07 39.5	Kurile Islands 42.96 N 146.48 E
	e	AB	07 49	H = 18 55 39.6 h = 45 km MB=5.7 MS=5.4
	eS	B	17 38	D = 78.82 Az = 332.6 (NEIS)
	LmH	B	46.0	PV A 1.8s 121.6nm M = 5.6
	LmV	B	46.9	LmH B 17 3.8/um 5.8
				LmV B 16 2.9/um 5.7

Day	Phase		h m s	Remarks
17.	eP	A	19 15 38	Off Coast of Hokkaido, Japan
	eS	B	25 36	42.68 N 146.34 E
				H = 19 03 35.1 h = 32 km MB=5.6 MS=5.1
				D = 79.02 Az = 332.6 (NEIS)
				PV A 1.5s 35.2nm M = 5.3
17.	-eiP	A	20 49 57	Off Coast of Hokkaido, Japan
	Pm	A	50 13	42.71 N 146.04 E
	eS	B	21 00 00	H = 20 37 57.3 h = 50 km MB=6.0 MS=6.0
	LmH	B	28.5	D = 78.90 Az = 332.4 (NEIS)
	LmV	B	29.3	PV A 2.1s 171.0nm M = 5.7
				PV B 7 1.8 µm 6.2
				PmV A 1.5 404.0nm 6.2
				SH B 13 4.4 µm 6.4
				LmH B 18 17.5 µm 6.4
				LmV B 16 15.7 µm 6.5
17.	e	A	21 15 47	Off Coast of Hokkaido, Japan
				42.84 N 145.95 E
				H = 21 03 35.7 h = 47 km MB = 5.0
				D = 78.75 Az = 332.3 (NEIS)
17.	eP	A	21 35 43	Hokkaido, Japan Region 42.96 N 145.95 E
				H = 21 23 42.6 h = normal MB = 4.8
				D = 78.64 Az = 332.3 (NEIS)
17.	eP	A	21 38 11.5	Hokkaido, Japan Region 43.00 N 145.34 E
	e	A	38 27.5	H = 21 26 14.2 h = 50 km MB = 5.2
				D = 78.40 Az = 332.0 (NEIS)
				PV A 1.0s 17.7nm M = 5.0
17.	eP	A	22 27 30.5	Hokkaido, Japan Region 42.97 N 145.50 E
				H = 22 15 33.2 h = 50 km MB = 4.9
				D = 78.48 Az = 332.1 (NEIS)
18.	eP	A	02 08 36	Hokkaido, Japan Region 41.07 N 143.71 E
				H = 01 56 30.9 h = 34 km MB = 4.5
				D = 79.52 Az = 331.3 (NEIS)

Day	Phase		h m s	Remarks
18.	eP	A	02 24 09.5	Off Coast of Hokkaido, Japan
	e	A	24 20	42.76 N 146.78 E
				H = 02 12 06.9 h = 36 km MB=4.9 MS=4.7
				D = 79.10 Az = 332.8 (NEIS)
18.	eP	A	02 31 33	Off Coast of Hokkaido, Japan
	e	A	31 41.5	42.60 N 146.25 E
	e	A	31 52.5	H = 02 19 30.0 h = 30 km MB=5.2 MS=4.8
	LmH	B	03 03.7	D = 79.06 Az = 332.5 (NEIS)
	LmV	B	06.6	LmH B 16s 1.35 μm M = 5.4
				LmV B 14 1.0 μm 5.4
18.	ePKHKP	A	03 16 49	Fiji Islands Region 21.11 S 179.31 W
	ePKP2	A	16 56.5	H = 02 58 10.1 h = 648 km MB = 4.9
				D = 149.29 Az = 346.3 (NEIS)
18.	eP	A	05 49 40	Off Coast of Hokkaido, Japan
	e	A	49 53	42.60 N 146.38 E
				H = 05 37 37.7 h = 38 km MB=5.5 MS=5.7
				D = 79.10 Az = 332.6 (NEIS)
				PV A 1.7s 33.3nm M = 5.0
18.	-eP	A	10 29 23.5	South of Alaska 52.18 N 164.87 W
				H = 10 17 26.3 h = 15 km MB=5.4 MS=4.5
				D = 77.51 Az = 2.3 (NEIS)
				PV A 1.8s 94.6nm M = 5.6
18.	eP	A	14 43 41	Kurile Islands 43.42 N 147.13 E
				H = 14 31 39.8 h = 36 km MB = 4.7
				D = 78.63 Az = 332.9 (NEIS)
				PV A 1.4s 14.0nm M = 4.8
18.	eP1	AB	17 57 46.5	Off Coast of Hokkaido, Japan
	eP2	AB	57 56.5	42.50 N 145.97 E
	eS	B	18 07 40	H = 17 45 43.7 h = 29 km MB=5.8 MS=5.5
	LmH	B	29.8	D = 79.06 Az = 332.4 (NEIS)
	LmV	B	38.9	P1V A 2.0s 94.0nm M = 5.5
				P2V A 1.7 172.5nm 5.8
				LmH B 21 8.5 μm 6.1
				LmV B 14 2.6 μm 5.7

Day	Phase		h m s	Remarks
18.	eP	A	18 36 22.5	Hokkaido, Japan Region 42.31 N 145.41 E
	e	A	36 26	H = 18 24 19.6 h = 29 km MB = 5.3
	e	A	36 35.5	D = 79.03 Az = 332.1 (NEIS)
18.	eP	A	20 43 40	Off Coast of Hokkaido, Japan
	e	A	43 54	42.71 N 146.29 E
				H = 20 31 37.4 h = 34 km MB = 5.1
				D = 78.98 Az = 332.5 (NEIS)
				PV A traces
19.	eP	A	02 34 05	Off Coast of Hokkaido, Japan
	LmV	B	03 13.0	42.91 N 146.72 E
	LmH	B	13.4	H = 02 22 05.6 h = 49 km MB=5.2 MS=4.9
				D = 78.95 Az = 332.8 (NEIS)
				LmH B 15s 1.0/um M = 5.3
				LmV B 17 0.9/um 5.2
19.	eP	A	03 06 10	Off Coast of Hokkaido, Japan
	e	A	06 22	42.72 N 146.04 E
				H = 02 54 09.8 h = 43 km MB = 5.6
				D = 78.88 Az = 332.4 (NEIS)
				PV A 0.7s 19.2nm M = 5.2
19.	eP	A	03 48 27	West Caroline Islands 8.13 N 137.33 E
	ePKIKP	A	52 38	H = 03 34 19.4 h = normal MB=5.5 MS=5.8
	LmV	B	04 42.4	D = 105.05 Az = 327.6 (NEIS)
	LmH	B	42.5	LmH B 17.5s 9.6/um M = 6.4
				LmV B 17.5 11.6/um 6.5
19.	eP	A	04 41 29	Hokkaido, Japan Region 42.38 N 145.80 E
				H = 04 29 27.8 h = 52 km MB = 4.9
				D = 79.11 Az = 332.3 (NEIS)
				PV A 1.8s 16.9nm M = 4.7
19.	eP	A	06 38 13	Hokkaido, Japan Region 42.73 N 145.62 E
	e	A	38 24	H = 06 26 12.8 h = normal MB = 4.6
				D = 78.73 Az = 332.2 (NEIS)
				PV A traces

Day	Phase		h m s	Remarks
19.	eP	A	06 49 55	Off Coast of Hokkaido, Japan
	e	A	50 08	42.86 N 146.50 E
				H = 06 37 54.4 h = 42 km MB = 5.3
				D = 78.92 Az = 332.6 (NEIS)
				PV A 1.8s 33.8nm M = 5.0
19.	eP	A	08 48 20	Off Coast of Hokkaido, Japan
				42.92 N 146.48 E
				H = 08 36 18.9 h = 36 km MB=5.3
				MS = 4.5 (NEIS)
				D = 78.86
19.	LmH	B	09 17.4	LmH B 15s 0.55/um
	LmV	B	17.5	LmV B 17 0.5/um
19.	LmH	B	09 20.0	LmH B 24s 1.0/um
	LmV	B	25.6	LmV B 19 0.7/um
19.	eP	A	09 49 27	Off East Coast of Hokkaido, Japan
				42.91 N 146.46 E
				H = 09 37 19.3 h = 42 km MB = 5.1
				D = 78.85 Az = 332.6 (NEIS)
19.	eP	A	11 10 51	Off East Coast of Hokkaido, Japan
				42.93 N 146.46 E
				H = 10 58 49.2 h = 37 km MB = 5.0
				D = 78.84 Az = 332.6 (NEIS)
				traces
19.	ePKIKP	A	17 05 31	Fiji Islands Region 20.26 S 177.98 W
	e(PKHKP)	A	05 37	H = 16 46 16.6 h = 258 km MB = 4.4
				D = 148.74 Az = 348.2 (NEIS)
				PKIKPV A 1.2s 18.3nm
19.	eP	A	17 22 16	Hokkaido, Japan Region 42.56 N 145.86 E
	LmH	B	18 01.0	H = 17 10 14.9 h = 41 km MB = 5.3
	LmV	B	01.0	D = 78.96 Az = 332.3 (NEIS)
				PV A 1.3s 15.3nm M = 4.8
				LmH B 17 0.35/um 4.8
				LmV B 17 0.3/um 4.7

Day	Phase		h m s	Remarks
19.	eP	A	20 17 53	Off Coast of Hokkaido, Japan
	epP	A	18 06	42.78 N 146.33 E
	LmH	B	53.5	H = 20 05 52.7 h = 45 km MB = 5.3
	LmV	B	56.5	D = 78.93 Az = 332.5 (NEIS)
				h = 48 km
				PV A 1.8s 32.0nm M = 5.1
				pPV A 1.5 47.7nm
				LmH B 14 0.35/um 4.9
				LmV B 17 0.35/um 4.7
19.	+eP	A	22 43 35.5	Near East Coast of Honshu, Japan
				36.45 N 140.61 E
				H = 22 31 18.5 h = 65 km MB = 5.2
				D = 82.35 Az = 330.0 (NEIS)
				PV A 1.2s 32.5nm M = 5.1
20.	eP	A	09 32 34	Kurile Islands 43.13 N 146.00 E
				H = 09 20 34.3 h = 36 km MB = 5.0
				D = 78.51 Az = 332.3 (NEIS)
20.	ePKIKP	A	12 21 49	Kermadec Islands Region 28.52 S 176.76 W
	ePKHKP	A	22 00	H = 12 01 56.7 h = 41 km MB= 5.6
	ePKP2	A	22 19	MS = 5.6 (NEIS)
	ePP	B	25 55	D = 157.01
	eSS	B	45 48	PPV B 9s 1.0/um M = 5.9
	LmH	B	13 32.9	LmH B 20 1.2/um 5.5
	LmV	B	40.5	LmV B 18 1.7/um 5.9
20.	eSg	A	14 47 36	Poland 50.27 N 18.90 E
				H = 14 45 03.5 M = 3.2 (WAR)
				D = 4.65
				SgV A 1.0s 13.8nm
21.	PKP	A	10 34 10	Fiji Islands Region 17.67 S 178.69 W
				H = 10 15 33.4 h = 587 km MB = 4.2
				D = 146.08 Az = 348.2 (NEIS)
				traces

Day	Phase		h m s	Remarks
21.	+eP	A	14 57 18.5	Southern Nevada 37.08 N 115.99 W
				H = 14 44 59.7 h = 5 km MB = 5.3
				D = 81.24 Az = 30.7 (NEIS)
				PV A 1.2s 26.4nm M = 5.2
22.	ePKIKP	A	02 19 07.5	Fiji Islands Region 17.63 S 178.88 W
	ePKHKP	A	19 09.5	H = 02 00 31.7 h = 565 km MB = 5.0
				D = 146.00 Az = 348.0 (NEIS)
				PKHKPV A 1.6s 46.7nm
22.	+eP	AB	06 19 36	Off Coast of Hokkaido, Japan
	eS	B	29 32	42.94 N 146.34 E
	LmH	B	57.8	H = 06 07 37.9 h = 53 km MB = 5.5
	LmV	B	58.1	D = 78.79 Az = 332.5 (NEIS)
				PV A 1.6s 71.4nm M = 5.4
				PV B 11 2.0 µm 6.0
				LmH B 17.5 19.6 µm
				LmV B 18 17.2 µm
22.	ePKP	A	17 36 23	Fiji Islands Region 17.01 S 175.80 E
				H = 17 16 47.5 h = 16 km MB=5.3 MS=4.9
				D = 144.12 Az = 342.8 (NEIS)
22.	eP	A	19 00 33	Mariana Islands Region 21.94 N 142.90 E
	eiPP	A	04 30	H = 18 47 36.6 h = 269 km MB = 5.2
				D = 95.94 Az = 331.2 (NEIS)
				PV A 1.2s 14.2nm M = 5.0
				PPV A 1.1 36.3nm 5.4
22.	ePKIKP	A	20 11 25	Tonga Islands 21.17 S 174.26 W
	ePKHKP	A	11 29	H = 19 51 40.5 h = normal MB=5.5 MS=5.1
	ePKP2	A	11 35.5	D = 150.22 Az = 352.5 (NEIS)
	LmH	B	21 24.0	PKHKPV A 1.5s 70.4nm
	LmV	B	24.5	LmH B 18 0.35 µm M = 5.1
				LmV B 19 0.5 µm 5.3
23.	+eiP	A	02 21 40.5	Kurile Islands 43.18 N 147.33 E
	LmH	B	53.5	H = 02 09 40.1 h = 41 km MB=5.5 MS=4.9

Day	Phase		h m s	Remarks
cont. 23.	LmV	B	03 01.3	D = 78.91 Az = 333.1 (NEIS)
				PV A 1.4s 69.8nm M = 5.5
				LmH B 23 1.2/um 5.2
				LmV B 16 1.6/um 5.5
24.	+iP	A	02 55 22.5	Kurile Islands 43.32 N 146.44 E
	eS	B	03 05 15	H = 02 43 25.5 h = 50 km MB=6.3 MS=7.1
	eSS	B	11 28	D = 78.49 Az = 332.6 (NEIS)
	LmH	B	33.4	PV A 1.7s 1482.0nm M = 6.7
	LmV	B	34.4	PV B 20 83.5/um 7.4
				LmH B 17 344.0/um 7.8
				LmV B 17 384.0/um 7.8
24.	+eiP	A	03 16 16	Kurile Islands 43.20 N 146.79 E
				H = 03 04 18.6 h = 55 km MB = 5.8 (NEIS)
				D = 78.71
				PV A 1.8s 270.0nm M = 5.9
24.	eP	A	03 40 37	Kurile Islands 43.30 N 146.82 E
				H = 03 28 38.5 h = 47 km MB = 5.5
				D = 78.63 Az = 332.8 (NEIS)
				PV A 1.8s 142.0nm M = 5.7
24.	eP	A	03 52 36	Kurile Islands 43.24 N 146.74 E
				H = 03 40 35.0 h = normal MB = 5.1
				D = 78.66 Az = 332.7 (NEIS)
24.	eP	A	03 59 47	Kurile Islands 42.97 N 146.64 E
				H = 03 47 45.8 h = 37 km MB = 5.0
				D = 78.86 Az = 332.7 (NEIS)
24.	eP	A	04 55 23	Kurile Islands 43.11 N 146.75 E
				H = 04 43 25.7 h = 62 km MB = 5.0
				D = 78.78 Az = 332.7 (NEIS)
				PV A 1.2s 14.2nm M = 4.9

Day	Phase		h m s	Remarks
24.	+iP	A	05 19 45.5	Kurile Islands 43.13 N 146.57 E
				H = 05 07 46.8 h = 44 km MB = 5.8
				D = 78.70 Az = 332.7 (NEIS)
				PV A 1.8s 121.5nm M = 5.9
24.	eP	A	07 20 52	Kurile Islands 43.14 N 146.69 E
				H = 07 08 52.1 h = 42 km MB = 4.8
				D = 78.73 Az = 332.7 (NEIS)
24.	+eP	A	11 05 51.5	Kurile Islands 43.06 N 146.48 E
	epP	A	06 06.5	H = 10 53 54.1 h = 60 km MB = 5.4
	LmH	B	44.8	D = 78.73 Az = 332.6 (NEIS)
	LmV	B	44.8	h = 52 km
				PV A 1.3s 35.0nm M = 5.1
				LmH B 14 0.5/um
				LmV B 16 0.5/um
24.	ePP	A	17 32 35	Marianas Region 20.95 N 143.26 E
	LmH	B	18 15.5	H = 17 15 13 h = 51 km (ISC)
	LmV	B	16.0	D = 97.0
				PPV A 2.5s 46.1nm M = 5.5
				LmH B 16 0.5/um
				LmV B 20 0.7/um
24.	eP	A	18 53 50.5	Kurile Islands 43.06 N 146.95 E
	epP	A	54 01.5	H = 18 41 49.3 h = 36 km MB = 5.2
				D = 78.89 Az = 332.9 (NEIS)
				h = 40 km
				PV A 1.5s 22.6nm M = 4.9
24.	eP	A	20 12 14	Kurile Islands 43.30 N 146.80 E
	epP	A	12 24	H = 20 00 16.1 h = 51 km MB = 5.1
	LmH	B	50.0	D = 78.63 Az = 332.8 (NEIS)
	LmV	B	53.2	h = 37 km
				PV A 1.9s 56.8nm M = 5.2
				pPV A 1.5 42.7nm
				LmH B 17 0.6/um
				LmV B 16 0.8/um

Day	Phase		h m s	Remarks
24.	ePKHKP	A	21 55 34.5	Fiji Islands Region 20.72 S 178.80 W
				H = 21 36 54.2 h = 624 km MB = 4.2
				D = 149.03 Az = 347.1 (NEIS)
25.	+eP	A	07 22 52.5	Andreanof Islands, Aleutian Is.
				52.92 N 174.68 W
				H = 07 11 23.5 h = 207 km MB = 5.2
				D = 76.68 Az = 355.9 (NEIS)
				PV A 1.2s 28.5nm M = 4.9
25.	eP	AB	07 32 28	Philippine Islands Region
	epP	A	32 39	19.11 N 121.15 E
	eSKS	B	42 50	H = 07 19 46.3 h = 50 km MB = 5.7
	eS	B	43 05	D = 87.16 Az = 323.1 (NEIS)
	ePS	B	43 55	h = 39 km
	LmH	B	08 15.2	PV A 2.0s 64.0nm M = 5.5
	LmV	B	15.2	pPV A 1.8 54.0nm
				LmH B 17.5 1.9/um 5.5
				LmV B 16 1.9/um 5.6
25.	eP	A	10 35 58	Iran 30.00 N 50.46 E
				H = 10 29 02.6 h = 48 km MB = 5.0
				D = 35.54 Az = 316.6 (NEIS)
25.	e(P)	A	11 52 00	Molucca Passage 0.95 N 125.99 E
	ePP	A	56 17.5	H = 11 37 51.6 h = normal MB=5.7 MS=4.8
				D = 104.48 Az = 323.2 (NEIS)
				(P)V A 1.5s 20.4nm M = 5.8
				PPV A 1.4 20.9nm 5.5
25.	-eP	A	21 00 25.5	Mindanao, Philippine Islands
				6.93 N 126.78 E
				H = 20 46 46.3 h = 73 km MB = 5.2
				D = 100.16 Az = 324.2 (NEIS)
26.	eP	A	00 52 53.5	Gulf of Alaska 59.41 N 144.69 W
				H = 00 41 48.8 h = 15 km MB = 4.4
				D = 68.63 Az = 15.9 (NEIS)

Day	Phase		h m s	Remarks
26.	eP	A	07 37 14.5	Guatemala 14.29 N 90.71 W
				H = 07 24 28.1 h = 38 km MB=4.5 MS=3.7
				D = 86.71 Az = 38.5 (NEIS)
				PV A 1.8s 16.9nm M = 5.0
26.	eP	A	11 51 22.5	Kurile Islands 43.06 N 146.61 E
	LmH	B	12 30.5	H = 11 39 23.9 h = 57 km MB = 5.3
	LmV	B	30.5	D = 78.77 Az = 332.7 (NEIS)
				PV A 1.4s 32.5nm M = 5.1
				LmH B 15 0.6 µm
				LmV B 16 0.7 µm
26.	eP	A	12 30 07.5	Kurile Islands 43.10 N 146.01 E
				H = 12 18 08.9 h = 46 km MB = 5.3
				D = 78.53 Az = 332.3 (NEIS)
				PV A 1.3s 26.2nm M = 5.1
26.	+eiP	AB	18 14 25.5	Kurile Islands 43.02 N 147.11 E
	eS	B	24 15	H = 18 02 24.2 h = 39 km MB=5.6 MS=5.8
	LmH	B	51.3	D = 78.98 Az = 333.0 (NEIS)
	LmV	B	57.0	PV A 1.9s 136.5nm M = 5.6
				PV B 10 1.4 µm 5.9
				LmH B 15 10.1 µm 6.3
				LmV B 14 6.9 µm 6.2
26.	eP	A	19 09 48	Crete 34.34 N 26.13 E
	eX	A	09 53.5	H = 19 05 21.9 h = 38 km MB = 4.8
				D = 19.44 Az = 331.3 (NEIS)
				PV A 0.6s 13.4nm M = 4.1
				XV A 0.9 29.2nm
26.	+iP1	AB	22 43 58.7	Kurile Islands 43.18 N 146.56 E
	+iP2	A	44 06	H = 22 32 00.2 h = 50 km MB=5.8 MS=6.6
	eiS	B	53 52	D = 78.66 Az = 332.6 (NEIS)
	LmH	B	20.2	P1V A 1.2s 36.6nm M = 5.3
	LmV	B	22.7	P2V A 2.0 376.0nm 6.0
				PV B 18 8.6 µm 6.4
				SH B 17 12.5 µm 6.8
				LmH B 16 66.0 µm 7.1
				LmV B 18 68.0 µm 7.1

Day	Phase		h m s	Remarks
26.	eP	A	22 53 36	<u>Off Coast of Hokkaido, Japan</u>
	epP	A	53 49	42.85 N 146.74 E
				H = 22 41 34.6 h = 41 km MB = 5.5
				D = 79.01 Az = 332.8 (NEIS)
				h = 48 km
				PV A 1.3s 39.3nm M = 5.3
				pPV A 1.6 49.5nm
26.	+eP	A	22 57 32.5	<u>Kurile Islands</u> 43.00 N 146.65 E
	epP	A	57 45.5	H = 22 45 32.5 h = 44 km MB = 5.4
				D = 78.84 Az = 332.7 (NEIS)
				h = 47 km
				PV A 2.2s 207.0nm M = 5.7
				pPV A 1.6 110.0nm
26.	eP	A	23 09 15	<u>Off Coast of Hokkaido, Japan</u>
	epP	A	09 27	42.80 N 146.64 E
				H = 22 57 13.0 h = 44 km MB = 5.3
				D = 79.01 Az = 332.7 (NEIS)
				h = 44 km
				PV traces
				pPV A 1.1s 12.1nm
27.	eP	A	01 15 01.5	<u>Off Coast of Hokkaido, Japan</u>
	epP	A	15 13	42.76 N 146.62 E
				H = 01 02 59.2 h = 41 km MB = 5.1
				D = 79.04 Az = 332.7 (NEIS)
				h = 41 km
27.	eP	A	01 53 15	<u>Kurile Islands</u> 43.10 N 146.50 E
	epP	A	53 25.5	H = 01 41 16.1 h = 46 km MB=5.4 MS=4.7
				D = 78.70 Az = 332.6 (NEIS)
				h = 40 km
				PV A 1.3s 34.9nm M = 5.2
27.	eP	A	03 27 19.5	<u>Kurile Islands</u> 43.13 N 146.79 E
				H = 03 15 20.4 h = 45 km MB = 5.0
				D = 78.77 Az = 332.8 (NEIS)
				PV A 1.4s 13.9nm M = 4.7

Day	Phase		h m s	Remarks
27.	eP	A	03 54 40	<u>Hokkaido, Japan Region</u> 42.55 N 145.78 E H = 03 42 38.0 h = 38 km MB=5.2 MS=4.5 D = 78.94 Az = 332.3 (NEIS) PV A 1.4s 18.6nm M = 4.9
27.	eP	A	04 09 07.5	<u>Southern Sinkiang Prov., China</u> 40.56 N 89.53 E
	LmV	B	32.5	H = 03 59 51.0 h = normal MB = 4.8
	LmH	B	33.5	D = 53.06 Az = 308.8 (NEIS) Atmospheric nuclear explosion (AEC) PV A 1.3s 15.3nm M = 4.8 LmH B 15 1.8 µm 5.3 LmV B 16 1.8 µm 5.2
27.	eP	A	12 02 09	<u>Near East Coast of Kamchatka</u> 53.16 N 159.76 E H = 11 50 40.0 h = 40 km MB = 5.0 D = 73.14 Az = 339.4 (NEIS) PV A 1.1s 28.2nm M = 5.2
27.	ePKIKP	A	12 36 41	<u>Kermadec Islands</u> 30.57 S 178.04 W
	ePKHKP	A	36 51	H = 12 16 48.7 h = 42 km MB=5.3 MS=5.0
	ePKP2	A	37 17	D = 158.68 Az = 342.9 (NEIS) PKIKPV A 1.7s 27.3nm
27.	eP	A	13 19 39	<u>Southern Sinkiang Prov., China</u>
	LmH	B	42.9	40.57 N 79.21 E
	LmV	B	42.9	H = 13 11 11.0 h = normal MB = 5.0 D = 46.81 Az = 306.2 (NEIS) PV A 1.5s 27.6nm M = 5.0 LmH B 17 0.6 µm 4.7 LmV B 16 0.7 µm 4.8
27.	eP	A	20 34 28	<u>Hokkaido, Japan Region</u> 42.67 N 145.42 E H = 20 22 27.3 h = normal MB = 4.7 D = 78.71 Az = 332.1 (NEIS) traces

Day	Phase		h m s	Remarks
27.	ePKHKP	A	21 13 45.5	South of Fiji Islands 23.74 S 179.87 E
	ePKP2	A	13 56	H = 20 54 51.5 h = 539 km MB = 4.7
				D = 151.63 Az = 344.2 (NEIS)
				PKHKPV A 1.5s 12.5nm
28.	eP	A	01 53 05	Near East Coast Kamchatka
				56.15 N 162.64 E
				H = 01 41 44.6 h = 0 km MB = 4.7
				D = 70.89 Az = 341 (ISC)
				PV A 1.0s 13.8nm M = 4.9
28.	ePKIKP	A	10 47 27	Fiji Islands Region 18.16 S 178.59 W
	ePKHKP	A	47 28.5	H = 10 28 56.5 h = 642 km MB = 4.5
				D = 146.58 Az = 348.2 (NEIS)
28.	eP	A	11 02 07	Southwestern Ryukyu Islands
	epP	AB	02 18	23.39 N 123.49 E
	eS	B	12 40	H = 10 49 33.4 h = 41 km MB=5.2 MS=5.2
	LmH	B	46.6	D = 85.04 Az = 323.6 (NEIS)
	LmV	B	48.8	h = 42 km
				PV A 1.4s 18.2nm M = 5.1
				pPV A 1.2 43.0nm
				LmH B 15.5 2.3 μm 5.7
				LmV B 16 1.6 μm 5.5
28.	ePKIKP	A	14 30 10.5	Santa Cruz Islands 12.79 S 166.69 E
	epPKIKP	A	30 45	H = 14 11 03.0 h = 146 km MB = 5.5
				D = 137.13 Az = 336.8 (NEIS)
				h = 127 km
				PKIKPV A 1.3s 14.1nm
28.	eP	A	19 27 31	Southern Nevada 37.15 N 116.09 W
				H = 19 15 12.4 h = 0 km MB = 4.9
				D = 81.22 Az = 30.6 (NEIS)
				Underground explosion "Portulaca" (AEC)
				traces

Day	Phase		h m s	Remarks
28.	eP	A	22 09 09.5	Szechwan Province, China
				29.09 N 103.59 E
				H = 21 58 02.9 h = normal MB = 5.2
				D = 69.28 Az = 317.1 (NEIS)
28.	eP	A	22 10 37	Szechwan Province, China
	eS	B	19 44	28.90 N 103.63 E
	LmH	B	37.6	H = 21 59 30.3 h = normal MB = 5.4
	LmV	B	42.8	D = 69.44 Az = 317.2 (NEIS)
				PV A 1.3s 35.0nm M = 5.3
				LmH B 24 1.2/um 5.1
				LmV B 16 0.5/um 4.9
29.	eP	A	00 15 54	Szechwan Province, China
				28.87 N 103.75 E
				H = 00 04 46.8 h = normal MB = 5.1
				D = 69.53 Az = 317.2 (NEIS)
				traces
29.	eP	A	01 39 48	Kurile Islands 43.59 N 145.99 E
				H = 01 27 53.8 h = 58 km MB = 4.9
				D = 78.09 Az = 332.3 (NEIS)
				PV A 1.0s 11.8nm M = 4.8
29.	eP	A	02 36 50	Mariana Islands Region 21.11 N 143.11 E
	ePP	AB	40 47	H = 02 23 20.9 h = 24 km MB=5.7 MS=5.3
	e	A	41 30.5	D = 96.76 Az = 331.3 (NEIS)
	eSS	C	54 48	PV A 2.0s 42.7nm M = 5.6
	LmH	B	03 23.5	PPV A 2.2 87.0nm 5.7
	LmV	B	23.8	LmH B 20 1.1/um 5.3
				LmV B 20 1.1/um 5.3
29.	+iP	AB	03 38 48.8	Hokkaido, Japan Region 43.40 N 145.80 E
	epP	A	39 04.5	H = 03 26 53.2 h = 50 km MB=5.9 MS=5.0
	esP	A	39 14	D = 78.20 Az = 332.2 (NEIS)
	e	A	39 24	h = 64 km
	eS	B	48 35	PV A 1.3s 179.0nm M = 5.9

Day	Phase		h m s	Remarks
cont. 29.	LmH	B	04 15.5	PV B 9s 0.8 µm M = 5.7
	LmV	B	16.8	SH B 8 0.6 µm 5.8
				LmH B 18 1.7 µm 5.4
				LmV B 18 1.3 µm 5.3
29.	LmH	B	08 42.0	Off Coast of Central America
	LmV	B	42.0	3.90 N 85.01 W
				H = 07 55 11.9 h = normal
				MB = 5.3 MS = 4.9 (NEIS)
				D = 91.2
				LmH B 22s 0.45 µm M = 4.9
				LmV B 22 0.45 µm 4.9
29.	eP	A	12 31 08	Kurile Islands 49.18 N 156.16 E
				H = 12 19 22.7 h = normal MB = 4.8
				D = 76.02 Az = 337.6 (NEIS)
				PV A 1.2s 24.4nm M = 5.1
29.	iPn	A	15 05 29.8	Dorheim/Hessen, Federal Republic of
	eSg	A	05 50.5	Germany, Explosion
				50°57.86' N 09°13.15' E
				H = 15 05 00.40 yield 5.4 t (Hannover)
				D = 1.5
29.	eP	A	18 03 07.5	Szechwan Province, China
	LmH	B	30.1	28.84 N 103.67 E
	LmV	B	35.2	H = 17 52 00.8 h = normal MB = 5.2
				D = 69.51 Az = 317.2 (NEIS)
				PV A 1.5s 25.2nm M = 5.1
				LmH B 22 0.8 µm 4.9
				LmV B 16 0.4 µm 4.8
29.	eP	A	23 20 07	North Atlantic Ocean 53.95 N 35.16 W
	LmH	B	31.3	H = 23 14 14.4 h = normal MB=4.6 MS=4.4
	LmV	B	31.3	D = 28.39 Az = 77.4 (NEIS)
				PV A 2.0s 42.7nm M = 4.8
				LmH B 16 0.4 µm 4.1
				LmV B 16 0.5 µm 4.4

Day	Phase		h m s	Remarks
30.	LmH	B	00 02.0	LmH B 15s 0.35/um
	LmV	B	02.3	LmV B 15 0.5/um
30.	LmV	B	01 15.5	LmH B 18s 0.2/um
	LmH	B	16.0	LmV B 19 0.35/um
30.	LmH	B	01 53.5	LmH B 15s 0.35/um
	LmV	B	55.8	LmV B 13 0.55/um
30.	ePKIKP	A	05 42 49	Fiji Islands Region 19.87 S 177.62 W
	eiPKHKP	A	42 53.5	H = 05 24 10.0 h = 576 km MB = 5.0
	e	A	43 05.5	D = 148.43 Az = 348.7 (NEIS)
				PKIKPV A 1.4s 11.6nm
				PKHKPV A 1.3 45.9nm
30.	LmV	B	09 03.0	Near Coast of Guatemala
	LmH	B	04.0	13.76 N 90.94 W
				H = 08 08 57.6 h = 78 km MB = 5.1 (NEIS)
				D = 87.2
				LmH B 17s 0.5/um
				LmV B 18 0.6/um
30.	ePKP	A	10 23 56	Tonga Islands Region 17.24 S 172.43 W
				H = 10 04 15.8 h = normal MB = 4.5
				D = 146.52 Az = 355.3 (NEIS)
				PKPV A 1.8s 16.9nm
30.	eP	A	17 55 58	Taiwan Region 22.86 N 121.38 E
				H = 17 43 27.6 h = normal MB = 5.2
				D = 84.30 Az = 323.0 (NEIS)
				PV A 1.3s 21.8nm M = 5.2
30.	eP	A	18 07 38	Near Islands, Aleutian Is. 52.75N 172.26E
	eS	C	17(45)	H = 17 55 55.9 h = 44 km MB=5.4 MS=4.8
	eSS	C	22 20	D = 75.69 Az = 347.4 (NEIS)
	LmH	C	38.5	PV A 1.6s 55.0nm M = 5.3
	LmV	C	38.5	LmH C 17 2.3/um 5.5
				LmV C 18 3.7/um 5.8

Day	Phase		h m s	Remarks
1.	eP	A	02 05 58	Sea of Okhotsk 48.38 N 145.95 E
				H = 01 55 08.0 h = 435 km MB = 4.4
				D = 73.85 Az = 331.7 (NEIS)
				traces
1.	+iP1	AB	13 44 38	Off Coast of Southeastern Alaska
	eP2	AB	44 48	57.84 N 137.33 W
	ePP	B	46 52	H = 13 33 34.6 h = normal MB=6.1 MS=6.7
	eS	B	53 44	D = 68.91 Az = 20.6 (NEIS)
	eSS	B	58 05	P1V A 1.8s 365.0nm M = 6.2
	eP'P'1	A	14 12 45	P1V B 8 15.3/um 7.1
	eP'P'2	A	12 55	P2V A 4.0 9900.0nm 7.2
	LmH	B	19.4	P2V B 13 21.8/um 7.1
	LmV	B	21.0	PPV B 13 10.5/um 6.9
				P'P'1V A 4.0 2220.0nm
				P'P'2V A 4.0 4280.0nm
				LmH B 14 38.9/um 6.8
				LmV B 14 41.5/um 6.9
1.	+eP	A	15 23 09	Off Coast of Southeastern Alaska
				57.78 N 137.29 W
				H = 15 12 05.0 h = normal MB = 5.2
				D = 68.96 Az = 20.6 (NEIS)
				PV A 1.0s 35.4nm M = 5.4
2.	eP	A	01 10 23	North Atlantic Ridge 49.52 N 28.52 W
	LmH	B	20.0	H = 01 04 56.0 h = normal MB = 5.0
	LmV	B	20.4	D = 25.56 Az = 72.0 (NEIS)
				PV A 1.6s 44.0nm M = 4.8
				LmH B 16 1.3/um 4.6
				LmV B 16 1.2/um 4.6
2.	eP	A	01 52 39	North of Svalbard 84.08 N 0.58 W
	LmH	C	02 07.5	H = 01 45 59.4 h = normal MB = 4.7
	LmV	C	07.5	D = 33.74 Az = 166.0 (NEIS)
				PV A 1.0s 11.8nm M = 4.8
				LmH C 19 0.45/um 4.2
				LmV C 19 0.3/um 4.2

Day	Phase		h m s	Remarks
2.	eP	A	06 03 27	Hokkaido, Japan Region 42.91 N 145.40 E
	epP	A	03 44	H = 05 51 30.9 h = 57 km MB = 5.2
	esP	A	03 51	D = 78.49 Az = 332.0 (NEIS)
				h = 63 km
				PV A 1.3s 21.8nm M = 5.0
2.	-eP	A	06 07 41.5	Komandorsky Islands Region
	e	A	07 51	54.04 N 164.09 E
	LmH	B	45.7	H = 05 56 12.4 h = normal MB = 5.4
	LmV	B	47.0	D = 73.16 Az = 342.1 (NEIS)
				PV A 1.2s 85.5nm M = 5.6
				LmH B 14 0.4/um 4.8
				LmV B 14 0.3/um 4.7
2.	LmH	B	12 22.2	Aegean Sea 39.68 N 23.73 E
	LmV	B	25.1	H = 12 14 09.8 h = 33 km MB = 4.1 (ISC)
				D = 14.0
				LmH B 10.5s 0.9/um M = 4.1
				LmV B 10 0.55/um
2.	eP	A	13 47 54.5	Southeastern Alaska 57.94 N 136.93 W
				H = 13 36 52.6 h = normal MB = 3.8
				D = 68.74 Az = 20.9 (NEIS)
2.	eP	A	18 33 38.5	Dominican Republic Region
				17.20 N 71.68 W
				H = 18 22 10.4 h = 25 km MB = 4.8
				D = 72.68 Az = 41.2 (NEIS)
				PV A 1.4s 16.3nm M = 4.9
2.	ePKP2	A	20 15 50.5	Kermadec Islands 29.73 S 177.63 W
	LmH	C	21 20.0	H = 19 55 26.0 h = 51 km MB = 4.8 (NEIS)
	LmV	C	20.0	D = 158.0
				PKP2V A 1.9s 34.1nm
				LmH C 25 0.25/um
				LmV C 25 0.3/um

Day	Phase		h m s	Remarks
3.	eP	AB	04 12 37.5	Michoacan, Mexico 19.10 N 101.83 W
	e	AC	13 01.5	H = 03 59 53.7 h = 125 km MB = 5.6
	e	A	13 17	D = 89.35 Az = 35.7 (NEIS)
	e	A	13 38.5	PV A 1.6s 41.2nm M = 5.3
	ePP	B	16 08	LmH B 32 2.2/um
	eSKS	C	23 00	LmV B 32 3.0/um
	esS	C	24 00	
	ePS	C	24 20	
	LmH	B	45.0	
	LmV	B	45.0	
3.	eP	AB	06 50 54	Samar, Philippine Islands
	ePP	AB	54 44.5	12.27 N 125.43 E
	e	B	07 00 40	H = 06 37 34.4 h = 44 km MB=5.5 MS=6.0
	eS	B	02 08	D = 95.08 Az = 324.2 (NEIS)
				PV A 1.0s 19.7nm M = 5.5
				PV B 15 0.5/um 5.8
				PPV B 15 0.6/um 5.8
				SH B 17 1.7/um 6.1
				Long waves superimposed by the following earthquake
3.	+eP	AB	07 17 05	Samar, Philippine Islands
	ePP	B	20 52	12.21 N 125.33 E
	ePPP	B	22 40	H = 07 03 43.9 h = normal MB=6.1 MS=6.5
	eSKS	B	27 35	D = 95.07 Az = 324.2 (NEIS)
	iS	BC	28 20	PV A 2.0s 265.0nm M = 6.3
	LmH	B	59.0	PV B 16 2.6/um 6.6
	LmV	B	08 04.0	PPV B 16 3.1/um 6.5
				SH B 17 6.4/um 6.7
				LmH B 18 20.5/um 6.6
				LmV B 18 19.8/um 6.7
3.	eP	A	08 02 41.5	Samar, Philippine Islands
				12.28 N 125.40 E
				H = 07 49 22.3 h = 51 km MB = 5.4
				D = 95.05 Az = 324.2 (NEIS)
				PV A 1.2s 12.2nm M = 5.2

Day	Phase		h m s	Remarks
3.	eP	A	08 10 43	<u>Samar, Philippine Islands</u>
	ePP	A	14 34	12.14 N 125.39 E
	epPP	A	14 47.5	H = 07 57 24.0 h = 56 km MB = 5.3
				D = 95.16 Az = 324.2 (NEIS)
				PV A 1.6s 27.0nm M = 5.4
				PPV A 1.2 16.3nm 5.4
3.	eP	A	11 09 15	<u>Kyushu, Japan</u> 31.06 N 130.20 E
				H = 10 57 07.5 h = 139 km MB = 5.0
				D = 82.27 Az = 325.7 (NEIS)
				PV A 1.1s 20.2nm M = 4.8
3.	ePn	A	16 11 49.5	<u>Adriatic Sea</u> 44.08 N 13.27 E
	eSn	A	13 06.5	H = 16 10 12.4 h = 47 km MB = 5.3
	LmH	B	15.4	D = 6.67 Az = 350.9 (NEIS)
	LmV	B	15.6	PnV A 1.1s 28.2nm M = 4.8
				LmH B 7.5 0.8/um 3.7
				LmV B 5 0.6/um
3.	eP	A	16 41 40.5	<u>Southeastern Alaska</u> 58.06 N 137.73 W
				H = 16 30 37.5 h = normal MB = 4.6
				D = 68.78 Az = 20.4 (NEIS)
				PV A 1.3s 10.9nm M = 4.8
3.	+iP1	ABC	17 10 38.4	<u>Southeastern Alaska</u> 57.90 N 138.02 W
	iP2	A	10 42	H = 16 59 35.1 h = normal MB=6.0 MS=6.0
	ePP	BC	13 08	D = 68.91 Az = 20.2 (NEIS)
	eS	BC	19 45	P1V A 2.0s 308.0nm M = 6.0
	e	BC	19 54	P2V A 1.8 393.0nm 6.2
	eSS	C	24 16	SH B 13.5 2.4/um 5.9
	eP'P'1	AB	38 44	P'P'1V A 1.9 53.0nm
	eP'P'2	A	38 49	P'P'2V A 2.6 261.0nm
	eP'P'3	A	38 54.5	P'P'3V A 3.2 840.0nm
	LmV	B	41.4	LmH B 20 9.0/um 6.0
	LmH	B	41.5	LmV B 20 10.9/um 6.1

Day	Phase	h m s	Remarks
3.	eP A	17 55 20	Southeastern Alaska 57.99 N 137.88 W
			H = 17 44 16.5 h = normal MB = 5.1
			D = 68.87 Az = 20.3 (NEIS)
			PV A 1.2s 24.4nm M = 5.1
4.	eP A	10 02 18	Fox Islands, Aleutian Is.
	e A	03 44	52.07 N 171.36 W
			H = 09 50 24.0 h = 41 km MB = 4.7 MS=4.6
			D = 77.63 Az = 358.1 (NEIS)
			PV A 1.0s 19.7nm M = 5.1
4.	eP A	16 54 44	India-China Border Region
			27.24 N 92.47 E
			H = 16 44 11.5 h = 26 km MB = 5.2
			D = 63.82 Az = 315.5 (NEIS)
			PV A 1.0s 15.8nm M = 5.1
4.	eP A	21 15 33	Burma-India Border Region
	epP A	16 06	23.61 N 94.84 E
			H = 21 04 46.3 h = 127 km MB = 5.2
			D = 67.91 Az = 317.0 (NEIS)
			h = 139 km
			PV A 1.2s 16.3nm M = 4.8
4.	eP A	21 41 29.5	Off Coast of Hokkaido, Japan
			42.54 N 145.96 E
			H = 21 29 22.5 h = 36 km MB = 4.9
			D = 79.01 Az = 332.4 (NEIS)
4.	ePP1 AB	24 07 00	South of Africa 53.16 S 22.69 E
	ePP2 AB	07 05	H = 23 48 43.0 h = normal MB=5.5 MS=5.8
	e C	16 15	(NEIS)
	ei B	21 32	D = 103.85
	LmH B	49.9	PP1V A 2.0s 30.0nm M = 5.5
	LmV B	49.9	PP2V A 2.0 51.2nm 5.7
			LmH B 20 1.5/um 5.5
			LmV B 21 1.3/um 5.5

Day	Phase		h m s	Remarks
5.	eP	A	01 00 13	Central Mid-Atlantic Ridge 0 S 26 W
				H = 00 50 13 MB = 4.1 (NORSAR)
				D = 59.8
5.	+eP	AB	01 10 43	Kurile Islands 43.99 N 147.85 E
	eS	B	20 32	H = 00 58 45.0 h = 40 km MB = 5.5
	LmV	B	49.2	D = 78.36 Az = 333.3 (NEIS)
	LmH	B	50.0	PV A 1.7s 97.0nm M = 5.7
				LmH B 15 1.0/um 5.3
				LmV B 16 0.7/um 5.1
5.	ePKP	A	03 44 05	New Hebrides Islands 18.94 S 169.43 E
				H = 03 25 00.4 h = 251 km MB = 4.9
				D = 143.79 Az = 336.0 (NEIS)
				traces
5.	eP	A	07 56 28	Southeastern Alaska 58.02 N 137.30 W
	ePcP	A	56 50	H = 07 45 24.9 h = normal MB = 4.1
				D = 68.74 Az = 20.7 (NEIS)
				PV A 1.0s 9.9nm M = 4.8
5.	+eP	AB	08 00 08.5	Off Coast of Southeastern Alaska
	eS	BC	09 20	57.91 N 137.90 W
	eSS	C	13 45	H = 07 49 04.5 h = normal MB=5.4 MS=4.9
	eP'P'	A	28 16	D = 68.96 Az = 20.2 (NEIS)
	LmH	B	30.7	PV A 1.4s 56.0nm M = 5.4
	LmV	B	30.7	PV B 8 0.54/um 5.7
				P'P'V A 2.0 42.7nm
				LmH B 22 1.0/um 5.0
				LmV B 22 1.7/um 5.3
5.	LmH	B	14 42.7	Near Coast of Northern Chile
	LmV	B	42.7	27.18 S 71.01 W
				H = 13 37 11.1 h = 33 km MB = 5.1 (ISC)
				D = 106.2
				LmH B 17s 0.35/um M = 5.0
				LmV B 17 0.35/um 5.0

Day	Phase		h m s	Remarks
5.	eP	A	22 23 51	Albania 41.64 N 19.89 E
				H = 22 21 15.6 h = normal MB = 4.2
				D = 10.67 Az = 330.3 (NEIS)
5.	eP1	A	22 59 32	Luzon, Philippine Islands
	eP2	AB	59 35	13.23 N 124.70 E
	eP3	A	59 44.5	H = 22 46 16.4 h = 38 km MB=5.6 MS=6.0
	e	A	23 00 14.5	D = 93.89 Az = 324.1 (NEIS)
	e	A	03 02	P2V A 1.5s 40.2nm M = 5.6
	ePP	BC	03 32	P3V A 1.8 88.0nm 5.8
	ePPP	BC	05 22	PPV B 12 1.0/um 6.1
	eSKS	BC	10 00	LmH C 24 27.6/um 6.6
	eS	C	10 40	LmV B 16 18.8/um 6.7
	eiPS	BC	11 48	
	eSS	C	17 00	
	LmH	C	38.2	
	LmV	B	49.8	
6.	eP	C	09 41 40	Near Coast of Northern Chile
	ePP	BC	46 00	27.24 S 71.14 W
	eSKS	BC	52 30	H = 09 27 30.7 h = 34 km MB=5.4 MS=5.6
	eS	C	53 45	D = 106.23 Az = 41.1 (NEIS)
	ePS	BC	55 20	LmH B 18s 2.0/um M = 5.7
	eSS	C	10 01 00	LmV B 17 2.0/um 5.9
	LmV	B	33.8	
	LmH	B	34.1	
6.	eP	A	13 49 58.5	Near East Coast of Honshu, Japan
	epP	A	50 11	40.19 N 142.54 E
	LmH	B	14 29.7	H = 13 37 53.7 h = 51 km MB = 5.1
	LmV	B	29.7	D = 79.85 Az = 330.7 (NEIS)
				h = 46 km
				LmH B 15s 0.4/um M = 4.9
				LmV B 17 0.5/um 5.0
7.	ePKHKP	A	00 27 37	Fiji Islands Region 15.15 S 175.04 E
	ePKIKP	A	27 40	H = 00 09 15.5 h = 614 km MB = 5.2
				D = 142.14 Az = 342.8 (NEIS)

Day	Phase		h m s	Remarks
7.	+iPKP	A	02 46 47.9	New Hebrides Islands 19.08 S 168.82 E H = 02 27 24.8 h = 93 km MB = 5.6 D = 143.67 Az = 335.4 (NEIS) PKPV A 1.1s 40.4nm
7.	eP	A	03 09 11	Fox Islands, Aleutian Is. 52.49 N 168.26 W H = 02 57 19.2 h = 47 km MB = 4.8 D = 77.24 Az = 0.1 (NEIS) PV A 1.0s 19.7nm M = 5.1
7.	epP	A	12 58 56	Northern Chile 19.17 S 69.77 W H = 12 45 00.1 h = 89 km MB = 5.3 D = 99.30 Az = 39.6 (NEIS)
7.	ePKP	A	15 59 34.5	New Britain Region 6.02 S 150.60 E H = 15 40 40.0 h = 53 km MB = 5.4 D = 123.92 Az = 329.8 (NEIS)
7.	ePKP	AB	19 54 58	Easter Island Cordillera 50.01 S 113.91 W
	LmH	B	20 56.0	H = 19 35 19.1 h = normal MB = 4.5
	LmV	B	56.0	D = 145.88 Az = 67.5 (NEIS) LmH B 20s 0.7/um M = 5.4 LmV B 20 1.1/um 5.6
8.	eP	AB	00 16 06	Samar, Philippine Islands
	eSKS	C	26 35	12.29 N 125.63 E
	eS	BC	27 24	H = 00 02 46.4 h = 47 km MB = 5.3
	LmH	B	01 04.0	D = 95.17 Az = 324.3 (NEIS)
	LmV	B	04.0	PV A 1.6s 22.0nm M = 5.4 LmH B 16 1.3/um 5.4 LmV B 16 0.9/um 5.3
8.	ePn	A	03 59 34	Northern Italy 46.58 N 9.97 E
	ePg	A	59 50.5	H = 03 58 31.3 h = normal
	eSn	A	04 00 24	D = 4.21 Az = 14.4 (NEIS)
	eSg	A	00 46	

Day	Phase		h m s	Remarks
8.	eP	A	04 15 35	Northern Colombia 6.79 N 72.98 W
	epP	A	16 15.5	H = 04 03 34.5 h = 156 km MB = 5.7
				D = 81.37 Az = 39.9 (NEIS)
				h = 170 km
				PV A 1.2s 24.4nm M = 4.8
8.	eP	A	10 11 41.5	Kurile Islands 43.42 N 146.40 E
	e	A	11 52.5	H = 09 59 45.8 h = 61 km MB = 5.0
				D = 78.39 Az = 332.5 (NEIS)
				PV A 1.0s 19.7nm M = 5.0
8.	eSg	A	12 56 55	Poland 50.23 N 19.15 E
				H = 12 54 21.3 M = 3.0 (WAR)
				D = 4.8
8.	ePb	A	13 46 36.5	Northern Italy 44.39 N 9.79 E
				H = 13 44 31.7 h = normal
				D = 6.38 Az = 10.5 (NEIS)
8.	eP	A	19 05 50	Near East Coast of Kamchatka
				53.81 N 160.44 E
				H = 18 54 23.5 h = normal MB = 4.7
				D = 72.67 Az = 339.8 (NEIS)
				PV A 1.0s 9.8nm M = 4.8
8.	eSg	A	21 06 47	Poland 50.36 N 18.87 E
				H = 21 04 20.2 M = 2.9 (WAR)
				D = 4.7
8.	LmH	B	21 53.3	Near Coast of Northern Chile
	LmV	B	53.3	27.06 S 70.93 W
				H = 20 47 40 h = 21 km MB = 5.0 (ISC)
				D = 106.0
				LmH B 17s 0.55 µm M = 5.2
				LmV B 17 0.5 µm 5.1
9.	-iPn	A	00 28 03.5	Switzerland 46.81 N 9.77 E
	ePg	A	28 19.5	H = 00 27 02.9 h = 18 km

Day	Phase		h m s	Remarks
cont. 9.	iSn	A	00 28 52.2	D = 4.03 Az = 17.0 (NEIS)
	eiSg	AB	29 13	LmH B 8s 0.6 um M = 3.2
	LmH	B	29.7	LmV B 8 0.4 um
	LmV	B	29.8	
9.	eP	A	02 15 34	°<u>Kamchatka</u> 54.48 N 158.41 E
	e	A	15 38	H = 02 04 21.4 h = 100 km MB = 4.8
				D = 71.62 Az = 338.4 (NEIS)
				PV A 1.5s 32.6nm M = 4.9
9.	LmH	B	03 30.0	<u>Bismarck Sea</u> 3.36 S 149.02 E
	LmV	B	31.0	H = 02 18 56.3 h = 0 km (ISC)
				D = 120.8
				LmH B 20s 0.35 um M = 5.0
				LmV B 22 0.45 um 5.0
9.	+eiPKP	AB	11 22 01.5	<u>Tonga Islands</u> 16.61 S 174.83 W
				H = 11 02 49.6 h = 236 km MB = 5.0
				D = 145.66 Az = 352.7 (NEIS)
				PKPV A 1.3s 74.2nm
9.	ePg	A	15 05 31	<u>Lohne near Fritzlar, Fed. Rep. of Germany</u>
				51°11.36' N 9°14.66' E
				Explosion
				H = 15 05 00.70 yield 5.4 t (Hannover)
				D = 1.6
9.	+iP	AB	16 31 32	<u>Andaman Islands Region</u>
	epP	A	31 42.5	10.69 N 92.58 E
	esP	AB	31 46.5	H = 16 19 46.8 h = 46 km MB=5.7 MS=5.2
	ePP	BC	34 22	D = 76.10 Az = 319.6 (NEIS)
	e	BC	34 36	h = 41 km
	ePPP	C	36 20	LmH B 19s 0.8 um M = 5.1
	ePPPP	C	37 20	LmV B 20 0.9 um 5.1
	eS	BC	41 05	
	eSP	BC	41 40	
	LmH	B	17 09.6	
	LmV	B	09.6	

Day	Phase		h m s	Remarks
9.	ePKHKP	A	21 46 38.5	Fiji Islands Region 21.50 S 177.95 W
				H = 21 27 39.2 h = 463 km MB = 4.6
				D = 149.96 Az = 347.8 (NEIS)
				PKHKPV A 0.8s 15.4nm
10.	+iP	A	01 34 46.7	Eastern Kazakh SSR 49.78 N 78.06 E
	ePn	A	36 19	H = 01 26 57.6 h = 0 km MB = 5.4
				D = 41.23 Az = 297.7 (NEIS)
				Underground explosion MB = 5.7 (UPP)
				PV A 0.7s 72.8nm M = 5.5
10.	ePKIKP	AB	04 21 55	South of Fiji Islands 24.07 S 177.37 W
	ePKHKP	A	22 04	H = 04 02 18.5 h = 118 km MB = 5.1
	LmV	B	05 33.9	D = 152.57 Az = 347.5 (NEIS)
	LmH	B	39.0	PKIKPV A 3.0s 158.0nm
				PKHKPV A 3.0 237.0nm
				LmH B 18 1.2/um
				LmV B 20 1.1/um
10.	ePKIKP	AB	07 19 22.5	South of Fiji Islands 24.06 S 177.33 W
	ePKHKP	A	19 30	H = 06 59 43.7 h = 103 km MB = 5.2
	ePKP2	AB	19 39	D = 152.57 Az = 347.6 (NEIS)
	LmH	B	08 37.0	PKIKPV A 2.0s 42.7nm
	LmV	B	37.0	PKP2V A 1.8 121.6nm
				LmH B 19 1.5/um
				LmV B 19 1.4/um
10.	LmH	B	16 11.0	Near Coast of Northern Chile
	LmV	B	11.0	27.00 S 71.0 W
				H = 15 06 02.4 h = 34 km MB = 5.2 (ISC)
				D = 106.0
				LmH B 18s 0.65/um M = 5.2
				LmV B 18 0.7/um 5.2
10.	ePKP2	A	16 40 21.5	South of Fiji Islands 23.7 S 177.7 W
	e	A	40 29	H = 16 20 12 h = 0 km
				D = 152.16 Az = 347 (ISC)

Day	Phase		h m s	Remarks
10.	eP	AB	23 37 49	Off East Coast of Honshu, Japan
	ePP	A	41 03	37.52 N 142.49 E
	eS	BC	48 00	H = 23 25 31.2 h = 45 km MB = 5.2
	e(SS)	BC	54 32	D = 82.16 Az = 330.9 (NEIS)
	LmH	B	24 19.1	PV A 1.6s 35.7nm M = 5.2
	LmV	B	19.2	LmH B 14 1.4 µm 5.5
				LmV B 15 1.1 µm 5.4
11.	e	A	02 27 10.5	
	e	A	27 17.5	
	e	A	27 29	
	e	A	29 29	
11.	eP	A	05 28 31.5	Kurile Islands 43.24 N 146.35 E
				H = 05 16 34.7 h = 55 km MB = 4.4
				D = 78.52 Az = 332.5 (NEIS)
				traces
11.	eP	A	06 11 25	Molucca Sea 0.11 S 125.01 E
	ePP	A	15 47	H = 05 57 20.0 h = normal MB=5.4 MS=4.4
	LmH	B	07 00.7	D = 104.73 Az = 322.8 (NEIS)
	LmV	B	09.6	PV A 1.4s 14.0nm M = 5.7
11.	e(PKP2)	A	08 24 42	South of Fiji Islands 25.30 S 178.31 E
				H = 08 05 38.8 h = 627 km MB = 4.8
				D = 152.70 Az = 341.4 (NEIS)
				(PKP2)V A 1.3s 8.7nm
11.	eP	A	14 41 27	South of Honshu, Japan 30.82 N 142.10 E
	LmH	C	15 20.8	H = 14 28 40.6 h = 45 km MB = 5.1
	LmV	C	24.0	D = 87.85 Az = 331.0 (NEIS)
				PV A 1.4s 11.6nm M = 4.9
				LmH C 14 0.35 µm 4.9
				LmV C 16 0.4 µm 5.0
11.	eP	A	22 31 19.5	Taiwan 24.43 N 120.77 E
				H = 22 18 55 h = 29 km MB = 4.7
				D = 82.72 Az = 323 (ISC)
				PV A 1.6s 13.7nm M = 4.8

Day	Phase		h m s	Remarks
11.	ePKP	ABC	23 07 22	Tonga Islands 15.24 S 173.92 W
	LmH	B	24 14.3	H = 22 47 49.1 h = normal MB=5.4 MS=5.0
	LmV	B	15.8	D = 144.40 Az = 353.9 (NEIS)
				ePKPV A 2.1s 81.4nm
				LmH B 19 0.3 μm M = 5.1
				LmV B 18 0.3 μm 5.0
11.	eP	A	23 35 01.5	Andreanof Islands, Aleutian Is.
				51.97 N 176.10 W
				H = 23 23 11.7 h = 63 km MB = 5.1
				D = 77.55 Az = 355.0 (NEIS)
12.	eP	A	02 11 03	Chagos Archipelago Region
				5.31 S 68.62 E
				H = 01 59 28.7 h = normal MB = 4.9
				D = 74.07 Az = 326.3 (NEIS)
12.	eP	A	08 02 54	Near Islands, Aleutian Is.
	e	A	03 10	52.22 N 174.21 E
				H = 07 51 07.9 h = 47 km MB=5.2 MS=4.3
				D = 76.45 Az = 348.7 (NEIS)
				PV A 1.1s 20.2nm M = 5.0
12.	eP	A	08 08 17	Off Coast of Hokkaido, Japan
	eX	A	08 30.5	42.65 N 146.79 E
	LmH	B	47.0	H = 07 56 12.8 h = 21 km MB = 5.3
	LmV	B	47.0	D = 79.20 Az = 332.8 (NEIS)
				PV A 1.4s 27.9nm M = 5.1
				XV A 1.5 35.2
				LmH B 15.5 0.6 μm 5.1
				LmV B 16 0.4 μm 4.9
12.	LmV	B	14 50.0	Near Coast of Northern Chile
	LmH	B	51.0	27.07 S 71.15 W
				H = 13 45 28 h = 8 km MB = 5.4 (ISC)
				D = 106.2
				LmH B 18s 2.6 μm M = 5.8
				LmV B 18 2.2 μm 5.8

Day	Phase		h m s	Remarks
13.	e(PKP)	A	00 38 41	<u>Kermadec Islands Region</u> 27.51 S 176.77 W
	e	B	42 32	H = 00 18 34.4 h = 35 km MB = 4.4
	LmH	B	01 56.5	D = 156.03 Az = 346.8 (NEIS)
	LmV	B	57.0	LmH B 17s 0.3/um M = 5.1
				LmV B 18 0.25/um 5.1
13.	eP	A	03 11 18.5	<u>Vancouver Island Region</u>
	e	A	11 43	49.03 N 128.01 W
	LmH	B	43.7	H = 02 59 39.1 h = normal MB=5.3 MS=5.1
	LmV	B	43.9	D = 74.85 Az = 25.3 (NEIS)
				PV A 1.8s 33.8nm M = 5.0
				LmH B 18 0.3/um 4.7
				LmV B 16 0.4/um 4.9
13.	e	A	09 00 34	<u>Czechoslovakia</u> 50.60 N 14.16 E
				H = 09 00.1
				Explosion 8.3 t (KHC)
				D = 1.62 Az = 273 (ISC)
13.	LmH	C	15 11.8	<u>Solomon Islands</u> 6.32 S 154.82 E
	LmV	C	14.0	H = 14 06 08.8 h = 62 km MB = 5.2 (ISC)
				D = 125.3
				LmH C 22s 0.3/um
				LmV C 30 0.5/um
13.	eP	A	22 12 23.5	<u>Eastern Kashmir</u> 33.10 N 75.50 E
				H = 22 03 38.7 h = 52 km MB = 4.8
				D = 49.13 Az = 310.9 (NEIS)
14.	-eP	AB	05 00 49	<u>Tibet</u> 35.18 N 86.48 E
	ePP	B	02 54	H = 04 51 21.0 h = normal MB=6.0 MS=6.9
	eiS	B	08 25	D = 54.66 Az = 311.1 (NEIS)
	LmH	B	24.3	PV A 2.5s 1060.0nm M = 6.4
	LmV	B	26.4	PV B 12 4.3/um 6.4
				PPV B 14 4.4/um 6.3
				PPH B 15 3.3/um
				SH B 14 11.9/um 6.4
				LmH B 17 189.2/um 7.2
				LmV B 15 152.5/um 7.3

Day	Phase		h m s	Remarks
14.	eP	A	12 41 40	Southern Greece 37.90 N 21.07 E
	ePP	AB	41 49	H = 12 38 18.3 h = 36 km MB = 4.7
	e	A	42 04	D = 14.42 AZ = 335.2 (NEIS)
	LmH	B	48.3	LmH B 12s 2.4 μm M = 4.5
	LmV	B	48.3	LmV B 12 2.5 μm
14.	-iP	AB	13 48 57.8	Tibet 35.26 N 86.60 E
	ePP	B	51 00	H = 13 39 30.0 h = normal MB=5.9 MS=5.5
	eS	BC	56 32	D = 54.68 Az = 311.1 (NEIS)
	eSS	BC	14 00 25	PV A 2.0s 393.0nm M = 6.1
	LmH	B	12.1	PV B 5 1.7 μm 6.3
	LmV	B	14.6	PPV B 5 0.7 μm 5.9
				LmH B 14 6.1 μm 5.8
				LmV B 13 7.8 μm 6.0
15.	e(PKP)	A	13 06 35	Kermadec Islands Region 28.02 S 176.4 W
				H = 12 46 31.4 h = 69 km MB = 3.9
				D = 156.60 Az = 347.1 (NEIS)
15.	+eP	A	14 18 47.5	Kurile Islands 43.38 N 146.51 E
				H = 14 06 49.8 h = 43 km MB = 5.4
				D = 78.46 Az = 332.6 (NEIS)
				PV A 1.2s 36.6nm M = 5.3
15.	LmH	B	20 08.0	Kermadec Islands Region 28.06 S 176.34 W
	LmV	B	09.0	H = 18 30 19.2 h = 63 km MB = 4.6 (ISC)
				D = 156.7
				LmH B 17.5s 0.4 μm
				LmV B 18 0.4 μm
16.	ePKIKP	A	04 18 36.5	Dentrecasteaux Islands Region
	ePP	A	20 40	10.25 S 151.90 E
	LmV	B	05 16.6	H = 03 59 30.3 h = 40 km MB=5.3 MS=5.1
	LmH	B	16.7	D = 128.19 Az = 328.8 (NEIS)
				PPV A 2.0s 38.4nm M = 5.5
				LmH B 20 1.0 μm 5.5
				LmV B 21.5 1.7 μm 5.7

Day	Phase		h m s	Remarks
16.	e(P)	A	14 12 36	Near Coast of Libya 32.6 N 23.6 E
				H = 14 08 01 h = 57 km
				D = 20.08 Az = 337 (ISC)
16.	eP	ABC	18 25 54.5	Guerreco, Mexico 17.32 N 100.68 W
	e	A	26 25.5	H = 18 12 57.5 h = 44 km MB=5.6 MS=5.7
	ePP	ABC	29 28	D = 90.14 Az = 36.1 (NEIS)
	eS	BC	36 36	PV A 2.6s 147.0nm M = 5.8
	eSS	C	42 32	LmH B 17 2.5/um 5.7
	e(P'P')	C	51 36	LmV B 17 2.8/um 5.8
	LmH	B	19 10.0	
	LmV	B	11.2	
16.	eP	A	19 55 09	Tibet 35.07 N 86.45 E
				H = 19 45 38.3 h = 15 km MB = 5.4
				D = 54.71 Az = 311.2 (NEIS)
				PV A 1.8s 74.4nm M = 5.4
16.	eP	A	20 56 40	Near East Coast of Kamchatka
				55.08 N 161.00 E
				H = 20 45 26.4 h = 81 km MB = 5.0
				D = 71.59 Az = 340.0 (NEIS)
17.	LmH	B	01 16.6	West Irian Region 4.34 S 134.50 E
	LmV	B	21.8	H = 00 14 48.1 h = 33 km (ISC)
				D = 113.7
				LmH B 20.5s 0.5/um M = 5.1
				LmV B 20 0.4/um 5.0
17.	eP	A	09 33 52	Ionian Sea 37.72 N 19.82 E
				H = 09 30 29.7 h = normal MB = 4.0
				D = 14.19 Az = 338.2 (NEIS)
				traces
17.	ePKP	A	22 03 53	Loyalty Islands Region 23 S 171 E
				H = 21 44 04 MB = 4.1 (NORSAR)
				D = 148.0

Day	Phase		h m s	Remarks
18.	eP	A	04 06 34.5	<u>Luzon, Philippine Islands</u>
	LmV	B	49.2	14.93 N 119.86 E
	LmH	B	49.6	H = 03 53 41.2 h = 56 km MB = 5.1
				D = 89.73 Az = 322.8 (NEIS)
				PV A 1.2s 16.3nm M = 5.2
				LmH B 15.5 0.6/um
				LmV B 16 0.7/um
18.	eP	A	15 34 45	<u>Northern Chile</u> 18.35 S 69.20 W
	e	BC	45 12	H = 15 21 23.0 h = 151 km MB = 5.7
	LmV	B	16 16.8	D = 98.32 Az = 39.4 (NEIS)
	LmH	B	17.0	LmH B 22.5s 1.5/um
				LmV B 21 1.7/um
18.	ePKHKP	A	18 00 21	<u>Fiji Islands Region</u> 21.44 S 179.01 W
				H = 17 41 39.5 h = 633 km MB = 4.3
				D = 149.68 Az = 346.6 (NEIS)
				PKHKPV A 1.8s 20.3nm
19.	eP	A	02 33 10	<u>Kurile Islands</u> 43.51 N 146.77 E
				H = 02 21 14.8 h = 68 km MB = 4.2
				D = 78.43 Az = 332.7 (NEIS)
				traces
19.	eP	A	03 35 26	<u>Bay of Bengal</u> 7 N 89 E
				H = 03 23 38 MB = 4.1 (NORSAR)
				D = 76.6
19.	eP	A	04 18 17.5	<u>Kurile Islands</u> 43.23 N 146.36 E
				H = 04 06 21.6 h = 66 km MB = 4.8
				D = 78.54 Az = 332.5 (NEIS)
				traces
19.	ePKP	ABC	06 03 02	<u>Fiji Islands Region</u> 18.10 S 178.28 W
				H = 05 44 25.6 h = 571 km MB = 5.8
				D = 146.58 Az = 348.5 (NEIS)
				PKPV A 1.5s 201.0nm

Day	Phase		h m s	Remarks
19.	LmV	B	13 54.0	<u>Kurile Islands</u> 43.37 N 146.71 E
				H = 13 15 29.8 h = 46 km MB = 4.4 (ISC)
				D = 78.5
				LmV B 18s 0.35/um M = 4.7
19.	eP	A	19 33 47.5	<u>Greenland Sea</u> 73.78 N 8.57 E
	LmH	B	43.3	H = 19 28 43.7 h = normal MB = 4.7
	LmV	B	43.3	D = 23.26 Az = 175.1 (NEIS)
				PV A 1.4s 25.6nm M = 4.5
				LmH B 15 0.2/um 3.7
				LmV B 15 0.3/um 4.0
19.	ePKP	A	21 41 49	<u>Tonga Islands</u> 19.62 S 175.07 W
				H = 21 22 16.5 h = 133 km MB = 5.0
				D = 148.56 Az = 351.8 (NEIS)
				PKPV A 1.0s 11.8nm
20.	ePKHKP	A	08 21 11.5	<u>West of Maquarie Island</u>
				56.25 S 146.92 E
				H = 08 01 16.7 h = normal MB = 5.2
				D = 153.18 Az = 277.0 (NEIS)
				PKHKPV A 1.4s 23.2nm
20.	eP	ABC	08 25 12.5	<u>Near East Coast of Honshu, Japan</u>
	+i	A	25 31	36.41 N 140.95 E
	ePP	BC	28 16	H = 08 12 53.5 h = 46 km MB=5.8 MS=5.5
	eSKS	BC	35 26	D = 82.52 Az = 330.2 (NEIS)
	e(PS)	BC	35 50	PV A 2.4s 304.0nm M = 5.9
	LmH	B	09 06.1	PV B 8 1.58/um 6.1
	LmV	B	07.1	LmH B 17 4.9/um 5.9
				LmV B 15.5 6.5/um 6.1
20.	LmV	B	14 09.6	<u>Off Coast of Jalisco, Mexico</u>
	LmH	B	09.7	18.80 N 106.43 W
				H = 13 10 58 h = 5 km MB = 4.4 (ISC)
				D = 92.1
				LmH B 15s 0.5/um M = 5.1
				LmV B 15 0.8/um 5.4

Day	Phase		h m s	Remarks
20.	LmH	B	14 48.1	Off Coast of Jalisco, Mexico
	LmV	B	48.1	18.82 N 106.38 W
				H = 13 49 31 h = 41 km MB = 4.3 (ISC)
				D = 92.1
				LmH B 15s 0.6/um M = 5.2
				LmV B 15 0.8/um 5.4
20.	eP	A	20 01 35	Greenland Sea 79.75 N 4.68 E
				H = 19 55 33.6 h = normal MB = 4.3
				D = 29.32 Az = 171.0 (NEIS)
20.	eP	A	23 33 53.5	North of Svalbard 80.00 N 0.16 E
	e	A	33 57	H = 23 27 48.3 h = normal MB = 5.2
	e	A	34 01	D = 29.74 Az = 165.2 (NEIS)
	eS	BC	38 52	PV A 1.2s 32.5nm M = 5.0
	LmH	B	46.0	LmH B 17 0.5/um 4.2
	LmV	B	46.2	LmV B 16 0.5/um 4.3
20.	e	A	23 54 17.5	Kurile Islands 43.04 N 146.17 E
				H = 23 42 02.1 h = normal MB = 4.4
				D = 78.64 Az = 332.4 (NEIS)
				traces
21.	e	A	01 31 03	Hokkaido, Japan Region
	LmV	B	02 09.6	41.35 N 142.79 E
	LmH	B	09.7	H = 01 18 52.7 h = 34 km MB = 4.6
				D = 78.93 Az = 330.8 (NEIS)
				LmH B 16s 0.3/um M = 4.7
				LmV B 16 0.2/um 4.7
21.	ePKIKP	ABC	04 38 19	South of Fiji Islands 24.81 S 179.22 W
	iPKHKP	A	38 27.5	H = 04 19 17.1 h = 411 km MB = 5.9
	iPKP2	A	38 40.9	D = 152.88 Az = 344.8 (NEIS)
	e	BC	42 12	PKIKPV A 1.4s 177.0nm
	e	BC	44 30	PKHKPV A 1.3 332.0nm
	e	BC	05 01 04	PKP2V A 1.3 710.0nm
	e	BC	04 00	

Day	Phase		h m s	Remarks
21.	eP	A	05 35 31.5	<u>Tsinghai Province, China</u> 35.55 N 96.06 E H = 05 25 25.02 h = normal MB = 5.2 (NEIS) D = 60.22 PV A 1.3s 17.5nm
21.	ePKIKP	A	09 42 42	<u>New Britain Region</u> 6.20 S 150.71 E H = 09 23 49.1 h = 44 km MB = 4.7 (NEIS) D = 124.13
21.	eP	A	11 59 17.5	<u>Philippine Islands Region</u> 20.04 N 121.40 E H = 11 46 34.9 h = 72 km MB = 5.0 (NEIS) D = 86.56
21.	LmH LmV	B B	12 44.3 44.4	<u>Philippine Islands Region</u> 20.04 N 121.48 E H = 11 46 35.2 h = 72 km MB = 4.7 (ISC) D = 86.6 LmH B 16.3s 0.4 μm LmV B 16 0.6 μm
21.	eP	A	12 56 10.5	<u>Crete</u> 35.03 N 24.75 E H = 12 51 55.7 h = normal MB = 4.5 (NEIS) D = 18.29
21.	ePKHKP	A	19 32 20.5	<u>South of Fiji Islands</u> 22.06 S 179.37 W H = 19 13 36.2 h = 598 km MB = 4.9 (NEIS) D = 150.19
21.	eP LmH LmV	A B B	19 59 35 20 26.2 27.9	<u>Tibet</u> 35.11 N 86.42 E H = 19 50 06.3 h = normal MB = 5.3 (NEIS) D = 54.67 LmH B 14s 0.4 μm M = 4.6 LmV B 12 0.3 μm 4.7
22.	e e e	A BC BC	02 56 10.5 03 03 44 05 50	<u>South Sandwich Islands Region</u> 55.71 S 28.30 W H = 02 36 54.7 h = 70 km MB = 5.4 (NEIS)

Day	Phase		h m s	Remarks
cont.	LmH	B	03 36.4	D = 111.07
	LmV	B	36.5	LmH B 21.5s 0.7 µm
				LmV B 22 0.7 µm
22.	eP	A	05 42 38	Southern Honshu, Japan 34.25 N 136.87 E
				H = 05 30 55.8 h = 357 km MB = 3.7 (NEIS)
				D = 82.68
				traces
22.	ePKP2	A	15 31 27.5	Kermadec Islands Region
				30.52 S 179.33 W
				H = 15 11 25.6 h = 261 km MB = 4.5 (NEIS)
				D = 158.29
				PKP2V A 1.3s 34.9nm
22.	ePg	A	21 08 14	Vosges 48.3 N 6.5 E
	eSg	A	09 19	H = 21 06 58 (BCIS)
				D = 4.05
22.	ePKHKP	A	21 52 31	South of Fiji Islands 22.52 S 179.48 W
				H = 21 33 44.5 h = 595 km MB = 4.6 (NEIS)
				D = 150.62
23.	+iP	ABC	01 30 49	Eastern Kazakh SSR 49.99 N 78.85 E
	ePP	ABC	32 25.5	H = 01 22 57.8 h = 0 km MB=6.3 MS=4.7
	e	A	44 56	(NEIS)
	LmV	B	48.5	Underground explosion MB = 7.1 (UPP)
	LmH	B	48.6	D = 41.59
				PV A 1.2s 650.0nm M = 6.1
				LmH B 9.5 1.0 µm 5.0
				LmV B 12 1.1 µm 5.0
23.	LmH	B	09 45.3	West Irian 2.25 S 138.00 E
	LmV	B	45.5	H = 08 30 00 h = 21 km MB = 5.2 (ISC)
				D = 114.1
				LmH B 17s 0.2 µm M = 4.8
				LmV B 17.5 0.3 µm 5.0

Day	Phase	h m s	Remarks
23.	eP ABC	10 14 32	Taiwan Region 24.08 N 122.31 E
	eS BC	24 52	H = 10 02 05.2 h = 42 km MB = 5.4 (NEIS)
	LmH B	56.2	D = 83.85
	LmV B	56.2	PV A 1.6s 44.0nm M = 5.7
			LmH B 15 2.1/um 5.6
			LmV B 15 2.6/um 5.8
23.	ePKHKP A	14 43 52.5	South of Fiji 22.1 S 177.1 W
			H = 14 24 33 h = 280 km
			D = 150.74 Az = 349 (ISC)
			PKHKPV A 1.7s 18.2nm
23.	eP A	14 46 35	Taiwan Region 21.83 N 121.29 E
			H = 14 34 05.0 h = 73 km MB = 5.0 (NEIS)
			D = 85.08
			PV A 1.4s 41.8nm M = 5.3
23.	eP A	19 02 49.5	Andreanof Islands, Aleutian Is.
			52.31 N 174.75 W
			H = 18 51 02.3 h = 84 km MB = 4.7 (NEIS)
			D = 77.28
			PV A 1.2s 8.1nm M = 4.4
23.	eP A	20 50 25	Gulf of California 30.75 N 113.47 W
	LmH B	21 29.4	H = 20 37 46.1 h = normal MB = 5.0 (NEIS)
	LmV B	31.6	D = 85.57
			LmH B 15.5s 0.6/um M = 5.1
			LmV B 14 0.8/um 5.3
23.	ePn A	23 26 46.5	Northern Italy 44.14 N 12.69 E
	eSn A	27 59	H = 23 25 10.8 h = normal MB = 4.0 (NEIS)
	LmH B	30.2	D = 6.55
	LmV B	30.3	LmH B 12s 0.4/um M = 3.2
			LmV B 12 0.5/um
24.	ePn A	00 47 32.5	Austria 47.18 N 9.59 E
	iPg A	47 49.5	H = 00 46 38.1 h = normal (NEIS)
	eSn A	48 17.5	D = 3.72
	eSg A	48 37.5	

Day	Phase		h m s	Remarks
24.	eiPKP2	A	08 14 55	Kermadec Islands Region
	e	A	18 39.5	31.52 S 179.61 E
	LmV	B	09 10.6	H = 07 55 08.2 h = 429 km MB = 5.1 (NEIS)
	LmH	B	13.3	D = 158.91
				PKP2V A 1.6s 55.0nm
24.	ePP	A	13 16 09	Molucca Sea 0.01 S 124.90 E
	LmV	C	14 04.1	H = 12 57 48.0 h = 53 km MB = 5.4 (ISC)
	LmH	C	04.6	D = 104.6
				LmV C 19s 0.25/um
24.	eP	A	15 02 20	Kurile Islands 43.17 N 146.33 E
				H = 14 50 22.5 h = 58 km MB = 4.7 (NEIS)
				D = 78.59
				traces
24.	eP	A	20 14 15.5	South Atlantic Ocean 11.6 S 19.7 W
				H = 20 03 19 h = 33 km
				D = 67.68 Az = 21 (ISC)
24.	ePKP	A	20 21 58.5	Near Coast of Central Chile
	epPKP	A	22 13.5	30.53 S 71.60 W
	esPKP	A	22 25	H = 20 03 35.3 h = 60 km MB = 5.6 (NEIS)
	e	A	22 43.5	D = 108.96
	LmV	B	21 08.4	LmH B 18s 0.4/um
	LmH	B	08.7	LmV B 18 0.5/um
25.	ePKIKP	AB	06 27 48	Solomon Islands 8.68 S 160.73 E
	ePP	BC	30 00	H = 06 08 38.7 h = 69 km MB = 5.5
	e	A	30 04	D = 130.98 Az = 334.3 (NEIS)
	ePKS	BC	31 12	PKIKPV A 1.5s 20.1nm
	e	A	31 28	LmH B 18.8 3.1/um
	eSS	BC	47 24	LmV B 20 2.8/um
	LmH	B	07 19.0	
	LmV	B	25.4	
25.	ePKIKP	A	06 59 15	Solomon Islands 8.69 S 160.82 E
				H = 06 40 09.3 h = 77 km MB = 5.0
				D = 131.02 Az = 334.4 (NEIS)
				traces

Day	Phase		h m s	Remarks
25.	eP	A	20 12 18.5	Samar, Philippine Islands
	LmH	B	21 00.0	12.24 N 125.78 E
	LmV	B	00.2	H = 19 58 55.8 h = normal MB=5.3 MS=4.3
				D = 95.30 Az = 324.3 (NEIS)
				PV A 1.6s 27.5nm M = 5.4
				LmH B 15 0.5/um 5.1
				LmV B 16 0.6/um 5.2
26.	eP	A	04 35 23.5	Andaman Islands Region 10.46 N 93.28 E
				H = 04 23 36.3 h = 66 km MB = 4.4
				D = 76.71 Az = 319.7 (NEIS)
				traces
26.	LmV	B	07 41.2	Northern Chile 27.14 S 71.22 W
	LmH	B	41.6	H = 06 36 09.9 h = 31 km MB = 5.0 (ISC)
				D = 106.2
				LmV B 18s 0.4/um M = 5.0
26.	ePKIKP	A	11 14 19	New Ireland Region 5.12 S 153.31 E
				H = 10 55 24.8 h = 75 km MB = 4.6
				D = 124.47 Az = 331.4 (NEIS)
				PKIKPV A 1.4s 14.0nm
26.	ePKHKP	A	13 59 00	West of Macquarie Island
				52.80 S 140.04 E
				H = 13 39 15.8 h = normal MB = 5.6
				D = 148.54 Az = 287.1 (NEIS)
				PKHKPV A 1.4s 14.0nm
26.	ePg	A	15 05 20.5	Hilders/Rhön, Fed. Rep. of Germany
				50°32.55' N 10°02.31' E
				Explosion
				H = 15 05 00.62 yield 8.5 t (Hannover)
				D = 1.0
26.	eP	A	19 37 18	Andaman Islands Region 10.37 N 94.32 E
				H = 19 25 25.1 h = normal MB = 4.5
				D = 77.44 Az = 319.7 (NEIS)
				traces

Day	Phase		h m s	Remarks
26.	eP1	AB	20 18 26.5	Andaman Islands Region 10.47 N 93.89 E
	eP2	A	18 31.5	H = 20 06 33.2 h = normal MB = 5.1
	e	A	18 38	D = 77.09 Az = 319.7 (NEIS)
	LmH	B	56.3	P1V A 2.2s 60.0nm M = 5.2
	LmV	B	57.2	P2V A 2.0 85.5nm 5.4
				LmH B 19 0.4/um 4.8
				LmV B 20 0.6/um 4.9
27.	LmH	B	02 21.9	LmH B 18s 0.3/um
	LmV	B	22.2	LmV B 20 0.4/um
27.	ePn	A	10 03 03	Federal Republic of Germany
				51.74 N 6.80 E
				H = 10 02 09.1 h = 1 km
				D = 3.23 Az = 108.0 (NEIS)
27.	eP	A	15 04 38	Near East Coast of Honshu, Japan
				36.22 N 140.02 E
				H = 14 52 22.3 h = 81 km MB = 4.6
				D = 82.31 Az = 329.8 (NEIS)
				traces
27.	ePKP	A	16 28 05.5	New Hebrides Islands 17.22 S 167.85 E
				H = 16 08 32.6 h = 15 km MB=5.1 MS=4.9
				D = 141.61 Az = 335.6 (NEIS)
				traces
27.	ePP	A	17 05 49	South of Java 9.01 S 106.91 E
	LmV	B	37.4	H = 16 48 12.5 h = 56 km MB = 5.3 (ISC)
	LmH	B	38.4	D = 100.3
27.	eP	A	19 32 49.5	Kurile Islands 43.56 N 146.77 E
				H = 19 20 54.5 h = 62 km MB = 4.9
				D = 78.39 Az = 332.7 (NEIS)
27.	ePKP	ABC	19 46 16	Tonga Islands 15.54 S 173.09 W
	ePP	B	49 40	H = 19 26 41.9 h = normal MB=5.4 MS=5.6
	LmH	B	54.5	D = 144.78 Az = 354.8 (NEIS)

Day	Phase		h m s	Remarks
cont. 27.	LmV	B	19 58.1	PKPV A 1.4s 41.8nm
				PKPV B 12 1.62 µm
				LmH B 18 0.9 µm M = 5.6
				LmV B 18 1.3 µm 5.7
27.	eP	ABC	19 55 04.5	<u>Nicaragua</u> 12.81 N 86.67 W
	epP	ABC	55 52.5	H = 19 42 47.9 h = 199 km MB = 5.3
	e	ABC	58 32	D = 85.39 Az = 39.2 (NEIS)
	eS	BC	20 05 08	h = 200 km
	ePS	BC	06 40	PV A 2.8s 365.0nm M = 5.7
				pPV A 2.8 257.0nm
27.	eP	A	20 34 43	<u>Burma-India Border Region</u>
	ePcP	A	35 16	23.33 N 94.51 E
				H = 20 23 50.7 h = 79 km MB = 5.3
				D = 67.91 Az = 317.0 (NEIS)
				PV A 1.1s 34.3nm M = 5.2
28.	eP	A	03 53 16.5	<u>Off East Coast of Honshu, Japan</u>
	LmH	B	04 27.2	40.88 N 143.17 E
	LmV	B	28.3	H = 03 41 12.8 h = 38 km MB = 4.5
				D = 79.48 Az = 331.0 (NEIS)
				LmH B 18s 1.5 µm M = 5.4
				LmV B 18 0.4 µm 4.8
28.	ePKP2	A	04 21 47	<u>South of Kermadec Islands</u>
				32.38 S 178.17 W
				H = 04 01 09.9 h = normal MB = 4.9 (NEIS)
				D = 160.37
				PKP2V A 1.4s 27.9nm
28.	eP	ABC	14 40 48	<u>Off East Coast of Honshu, Japan</u>
	e	A	43 10	40.95 N 143.11 E
	eS	BC	50 44	H = 14 28 44.4 h = 37 km MB=5.3 MS=5.2
	LmH	B	15 14.7	D = 79.40 Az = 331.0 (NEIS)
	LmV	B	19.0	PV A 1.9s 79.5nm M = 5.4
				LmH B 17.5 2.3 µm 5.6
				LmV B 17 3.7 µm 5.8

Day	Phase		h m s	Remarks
28.	eP1	ABC	18 06 06	Taiwan Region 22.13 N 121.48 E
	iP2	A	06 10.7	H = 17 53 32.3 h = normal MB=5.4 MS=5.6
	ePP	BC	09 30	D = 84.93 Az = 323.0 (NEIS)
	eS	BC	16 32	P1V A 1.3s 39.3nm M = 5.4
	LmH	B	45.6	P2V A 2.0 162.0nm 5.9
	LmV	B	49.3	LmH B 14.5 4.0/um 5.9
				LmV B 16 6.3/um 6.3
28.	eP	A	18 59 44	Turkey 36.08 N 31.36 E
				H = 18 55 10.8 h = 70 km MB = 4.4
				D = 20.35 Az = 321.8 (NEIS)
				PV A 1.3s 43.6nm M = 4.6
28.	ePKHKP	A	19 34 34.5	Fiji Islands Region 18.87 S 178.20 W
				H = 19 15 46.4 h = 515 km MB = 4.8
				D = 147.35 Az = 348.4 (NEIS)
28.	eP	A	20 09 53.5	Southeastern Alaska 58.00 N 137.90 W
				H = 19 58 47.1 h = 10 km MB = 4.7
				D = 68.86 Az = 20.3 (NEIS)
				PV A 1.0s 11.8nm M = 5.0
28.	eiP	ABC	20 17 06.5	Sea of Okhotsk 50.48 N 148.85 E
	e	BC	17 12	H = 20 06 36.0 h = 592 km MB = 5.5
	epP	A	19 14	D = 72.87 Az = 333.1 (NEIS)
	ePP	B	20 00	h = 650 km
	eS	BC	25 46	PV A 1.8s 406.0nm M = 5.7
	esS	BC	29 15	PV B 8 1.25/um 5.5
	LmH	B	35.6	LmH B 17 0.8/um
	LmV	B	35.7	LmV B 16.5 0.7/um
28.	eP	A	21 00 57.5	Near Coast of Nicaragua 12.33 N 87.08 W
				H = 20 48 24.7 h = 79 km MB = 5.4
				D = 86.01 Az = 39.1 (NEIS)
28.	eP	A	22 31 40	Southern Peru 16.06 S 71.31 W
				H = 22 18 14.9 h = 110 km MB = 5.7
				D = 97.86 Az = 39.6 (NEIS)
				PV A 1.8s 27.0nm M = 5.5

Day	Phase		h m s	Remarks
29.	eP	A	01 24 23	Central Mid - Atlantic Ridge
				1.10 S 24.68 W
				H = 01 14 17.3 h = normal MB=5.2 MS=4.6
				D = 60.11 Az = 25.8 (NEIS)
				PV A 1.2s 12.2nm M = 4.9
29.	ePKHKP	A	01 45 53.5	Tonga 20.59 S 173.76 W
				H = 01 25 59.9 h = 0 km MB = 5.1
				D = 149.71 Az = 353 (ISC)
				PKHKPV A 1.8s 40.6nm
29.	eP	A	04 56 42.5	Kurile Islands 47.30 N 152.2 E
				H = 04 45 05.5 h = 137 km MB = 4.5
				D = 76.70 Az = 335 (ISC)
				PV A 1.0s 11.8nm M = 4.6
29.	eP	A	09 22 57.5	Kurile Islands 45.31 N 150.05 E
				H = 09 11 03.8 h = 46 km MB = 5.1
				D = 77.87 Az = 334.4 (NEIS)
				PV A 1.0s 15.7nm M = 5.0
29.	eP	ABC	15 03 01.5	Kurile Islands 43.09 N 146.80 E
	e	A	05 37	H = 14 51 03.1 h = 52 km MB = 5.4
	e	C	09 40	D = 78.81 Az = 332.8 (NEIS)
	LmH	B	41.2	PV A 1.4s 41.8nm M = 5.2
	LmV	B	42.1	LmH B 16.7 0.9 µm
				LmV B 16 0.9 µm
29.	eP	A	16 29 13.5	Peru 14.33 S 72.13 W
	ePP	ABC	33 14.5	H = 16 15 49.6 h = 82 km MB = 5.5
				D = 97.03 Az = 39.6 (NEIS)
				PV A 2.0s 29.9nm M = 5.5
				PPV A 2.6 95.3nm 5.9
29.	e(PKIKP)	A	22 02 13.5	West of Maquarie Island 56.32 S 147.39 E
	ePKHKP	A	02 18.5	H = 21 42 21.4 h = normal MB=5.3 MS=5.2
	e	A	02 31	D = 153.45 Az = 276.5 (NEIS)
	LmH	B	23 24.6	PKHKPV A 1.8s 37.2nm
	LmV	B	24.9	

Day	Phase		h m s	Remarks
30.	e	A	14 28 43	<u>Kyushu, Japan</u> 31.90 N 131.81 E H = 14 16 14.4 h = 41 km MB = 4.8 D = 82.35 Az = 326.3 (NEIS) LmH B 17s 0.5/um M = 4.9 LmV B 16 0.7/um 5.2
	LmV	B	15 09.2	
	LmH	B	09.3	
30.	ePKP	A	17 04 30	<u>Fiji Islands Region</u> 16.72 S 177.75 W H = 16 45 41.4 h = 445 km MB = 4.4 D = 145.33 Az = 349.5 (NEIS) PKPV A 1.0s 13.8nm
31.	ePKIKP	A	00 47 40	<u>East of North Island, N.Z.</u> 35.86 S 179.86 W H = 00 27 40.4 h = 58 km MB = 5.1 D = 163.06 Az = 334.2 (NEIS) PKIKPV A 1.8s 13.5nm
31.	LmV	B	06 50.5	<u>Near Coast of Central Chile</u> 37.66 S 73.31 W H = 05 41 04 h = 28 km MB = 5.1 (ISC) D = 84.9 LmH B 18.5s 3.9/um M = 5.8 LmV B 18 5.3/um 6.0
	LmH	B	50.8	
31.	eP dif	BC	11 05 24	<u>Near Coast of Northern Chile</u> 27.12 S 71.51 W H = 10 51 13.2 h = normal MB = 5.6 MS = 6.3 (NEIS) D = 106.37 LmH B 17.5s 9.3/um M = 6.4 LmV B 18 15.3/um 6.6
	e	A	08 45.5	
	ePP	ÁBC	09 45	
	eSKS	BC	16 16	
	e	B	17 28	
	ePS	BC	19 10	
	LmV	B	55.2	
	LmH	B	58.2	
31.	LmV	B	16 22.3	<u>Off Coast of Northern Chile</u> 26.88 S 71.17 W H = 15 17 31 h = 24 km MB = 4.8 (ISC) D = 106.0 or <u>Near Coast of Northern Chile</u>
	LmH	B	23.3	

Day	Phase		h m s	Remarks
cont. 31.				<u>Near Coast of Northern Chile</u>
				26.86 S 70.7 W
				H = 15 17 50 h = 45 km MB = 4.9 (ISC)
				D = 105.7
				LmH B 18s 1.0/um M = 5.4
				LmV B 18.5 1.2/um 5.5
31.	eP	A	17 03 33.5	<u>Near East Coast of Honshu, Japan</u>
				36.50 N 140.65 E
				H = 16 51 17.1 h = 61 km MB = 4.6
				D = 82.32 Az = 330.1 (NEIS)
31.	ePKIKP	ABC	21 04 08	<u>Solomon Islands</u> 8.77 S 160.99 E
	eSKP	BC	07 30	H = 20 44 52.2 h = 30 km MB=5.4 MS=6.0
	eSKS	BC	11 14	D = 131.17 Az = 334.5 (NEIS)
	eSS	BC	23 48	LmH B 19.5s 3.0/um M = 6.0
	LmH	B	55.0	LmV B 20.5 4.2/um 6.1
	LmV	B	22 02.0	
31.	eP	A	21 17 06.5	<u>Burma-India Border Region</u>
				27.26 N 97.09 E
				H = 21 06 14.9 h = normal MB = 4.8
				D = 66.67 Az = 316.3 (NEIS)
				PV A 1.4s 14.0nm M = 4.9

Day	Phase		h m s	Remarks
1.	ePKHKP	B	01 50 21	New Hebrides Islands 14.26 S 167.27 E
	ePKIKP	A	50 24.5	H = 01 31 30.9 h = 200 km MB = 6.1
	eX	A	50 27	D = 138.70 Az = 336.6 (NEIS)
	eY	A	50 32.5	PKIKPV A 1.2s 56.9nm
	ei	AB	50 34	XV A 2.6 88.4nm
	epPKP	B	51 13	YV A 2.4 2070.0nm
	ePP	B	53 20	
	ePP	A	53 25	
	iSKP	B	53 50	
	ei	B	54 11	
1.	LmV	B	12 13.5	Near Coast of Northern Chile
	LmH	B	14.3	Two shocks
				H1 = 11 08 24.4 27.15 S 71.05 W
				h = normal M = 5.2
				H2 = 11 13 37.9 27.12 S 71.18 W
				h = normal M = 5.2 (NEIS)
				D1 = 106.12
				D2 = 106.18
				LmH B 17s 0.5/um M = 5.1
				LmV B 17.5 0.6/um 5.2
1.	eP	A	14 15 17.5	Tibet 29.60 N 89.09 E
	e	A	15 43	H = 14 05 16.2 h = 71 km MB = 4.9
				D = 60.05 Az = 314.2 (NEIS)
1.	eP dif	BC	15 58 40	Off Coast of Northern Chile
	ePKP	A	16 02 52	26.83 S 71.00 W
	e	BC	09 22	H = 15 44 25.5 h = 16 km MB = 5.7 (NEIS)
	e	BC	12 18	D = 105.84
	LmV	B	48.8	LmH B 17s 1.6/um M = 5.6
	LmH	B	51.4	LmV B 18.5 2.1/um 5.7
1.	eP	A	18 08 40	Near East Coast of Honshu, Japan
				37.15 N 141.62 E
				H = 17 56 23.4 h = 48 km MB = 4.5
				D = 82.15 Az = 330.5 (NEIS)

Day	Phase		h m s	Remarks
1.	e	A	20 00 40.5	Turkey 40.89 N 34.60 E H = 19 56 08.5 h = 19 km MB = 4.3 (NEIS) D = 18.70
2.	eSg	A	00 54 40	Poland 50.35 N 18.87 E H = 00 52 12.4 M = 2.6 (WAR) D = 4.65
2.	e(Pg)	A	06 59 32	Czechoslovakia 48.73 N 14.11 E explosion yield 11.4 t (PRU) D = 2.5
2.	eP	A	07 24 15.5	Near East Coast of Kamchatka 52.60 N 158.79 E H = 07 12 44.5 h = normal MB = 4.6 D = 73.46 Az = 338.9 (NEIS)
2.	eP	ABC	09 09 30.5	Yunnan Province, China 27.81 N 104.54 E
	LmV	B	42.4	H = 08 58 15.1 h = normal MB = 5.4
	LmH	B	42.5	D = 70.78 Az = 317.7 (NEIS)
				PV A 1.6s 55.0nm M = 5.4
				LmH B 15.3 0.7 µm 5.1
				LmV B 16 1.0 µm 5.2
2.	LmH	B	10 46.2	Near Coast of Guerrero 16.60 N 98.19 W
	LmV	B	46.2	H = 10 09 00 h = 68 km MB = 4.4 (ISC)
				D = 89.3
				LmH B 16s 0.3 µm
				LmV B 17 0.5 µm
2.	eP	ABC	20 03 12	Iran 37.35 N 56.51 E
	LmH	B	19.2	H = 19 56 26.6 h = 36 km MB = 5.3
	LmV	B	20.2	D = 34.38 Az = 307.3 (NEIS)
				PV A 1.6s 93.4nm M = 5.5
				LmH B 14.5 0.5 µm 4.4
				LmV B 14.5 0.7 µm 4.6
2.	eP	A	20 35 43.5	Iran 37.33 N 56.52 E H = 20 28 55.7 h = 34 km MB = 4.8

Day	Phase		h m s	Remarks
cont. 2.				D = 34.40 Az = 307.3 (NEIS)
				PV A 1.3s 26.2nm M = 5.0
2.	LmH	C	22 37.5	<u>Taland Islands</u> 3.90 N 126.33 E
	LmV	C	42.0	H = 21 38 11 h = 35 km MB = 4.9 (ISC)
				D = 102.4
				LmV C 19s 0.3/um M = 4.9
3.	eP	A	04 08 45	<u>Fox Islands, Aleutian Is.</u>
				53.21 N 169.78 W
				H = 03 57 06.8 h = 124 km MB = 5.0
				D = 76.52 Az = 359.1 (NEIS)
				PV A 0.9s 37.0nm M = 5.2
3.	eP	A	06 38(24)	<u>Hokkaido, Japan Region</u> 42.50 N 145.75 E
	LmH	B	07 13.3	H = 06 26 25.4 h = 45 km MB=5.0 MS=4.5
	LmV	B	17.2	D = 78.97 Az = 332.3 (NEIS)
				LmH B 15s 0.4/um M = 4.9
				LmV B 17 0.35/um 4.8
3.	eP	A	15 55 46	<u>Haiti Region</u> 19.96 N 73.07 W
				H = 15 44 26.9 h = 37 km MB = 5.2
				D = 71.51 Az = 41.9 (NEIS)
				PV A 1.4s 46.5nm M = 5.3
3.	eP	ABC	17 35 21	<u>Near East Coast of Kamchatka</u>
	eS	BC	44 46	54.79 N 162.40 E
	LmH	B	18 10.1	H = 17 23 57.0 h = 28 km MB = 5.3
	LmV	B	13.1	D = 72.13 Az = 340.9 (NEIS)
				PV A 2.2s 175.0nm M = 5.7
				LmH B 16.5 2.4/um 5.6
				LmV B 16 1.9/um 5.5
3.	eP	A	17 43 08	<u>Near East Coast of Kamchatka</u>
				54.77 N 162.42 E
				H = 17 31 42.8 h = normal MB = 4.9
				D = 72.16 Az = 341.0 (NEIS)
				PV A 1.2s 12.2nm M = 4.8

Day	Phase		h m s	Remarks
3.	eP	A	17 46 44.5	Near East Coast of Kamchatka 54.74 N 162.36 E H = 17 35 20.4 h = normal MB = 4.9 D = 72.17 Az = 340.9 (NEIS) PV A 2.4s 82.8nm M = 5.3
3.	eP	A	17 57 47	Near East Coast of Kamchatka 54.63 N 161.96 E H = 17 46 22.7 h = normal MB = 4.8 D = 72.20 Az = 340.7 (NEIS)
3.	epP	A	19 25 20	Kurile Islands 43.01 N 147.64 E
	LmH	B	20 05.1	H = 19 13 05.1 h = 53 km MB = 4.6
	LmV	B	05.6	D = 79.16 Az = 333.2 (NEIS) h = 54 km LmH B 14s 0.7 μm LmV B 16 0.4 μm
3.	eP	A	23 30 43	Near East Coast of Kamchatka 54.82 N 162.39 E H = 23 19 18.8 h = normal MB = 4.7 D = 72.10 Az = 340.9 (NEIS)
4.	eP	ABC	00 57 23	Costa Rica 9.84 N 84.59 W
	e	A	57 35	H = 00 44 42.8 h = normal MB=5.3 MS=5.4
	eS	BC	01 07 46	D = 86.38 Az = 39.4 (NEIS)
	LmH	B	36.1	PV A 1.2s 22.4nm M = 5.3
	LmV	B	40.4	LmH B 19 1.2 μm 5.3 LmV B 16.5 1.1 μm 5.4
4.	eP	A	04 12 57	Greece 38.80 N 20.19 E H = 04 09 47.8 h = 39 km MB = 4.1 D = 13.31 Az = 335.7 (NEIS)
4.	eP	A	06 19 06	Hokkaido, Japan Region 42.84 N 145.47 E
	e	A	19 11	H = 06 07 08.5 h = 53 km MB = 4.9 D = 78.59 Az = 332.1 (NEIS)

Day	Phase		h m s	Remarks
4.	ePcP	A	07 40 07	Nicobar Islands Region 7.00 N 91.23 E
				H = 07 28 00.1 h = normal MB = 4.5
				D = 78.04 Az = 320.2 (NEIS)
				PcPV A 2.0s 25.6nm
4.	eP	A	20 02 43	Mid-Indian Rise 13.75 S 66.19 E
				H = 19 50 35.9 h = normal MB = 5.1
				D = 79.83 Az = 328.2 (NEIS)
				PV A 2.0s 25.6nm M = 4.9
4.	eP	A	22 05 28	Molucca Passage 1.99 N 126.72 E
	e	A	08 26	H = 21 51 25.0 h = 40 km MB=5.6 MS=5.2
	e	A	09 46	D = 104.08 Az = 323.5 (NEIS)
	eSKS	B	16 04	LmH B 15.5s 0.5/um M = 5.2
	LmH	B	58.5	LmV B 14.5 0.4/um 5.2
	LmV	B	58.6	
5.	LmH	B	01 45.5	Near Coast of Northern Chile
	LmV	B	46.3	22.81 S 70.22 W
				H = 00 49 58.3 h = 43 km MB = 5.5
				D = 102.35 Az = 40.2 (NEIS)
				PV A 2.0s 17.1nm M = 5.4
				LmV B 20 0.35/um 4.9
5.	LmH	C	08 51.0	Eastern New Guinea Region
	LmV	C	59.6	6.23 S 147.89 E
				H = 07 44 15.3 h = 62 km MB = 4.8 (ISC)
				D = 122.8
				LmH C 23s 0.5/um M = 5.1
				LmV C 20 0.6/um 5.3
5.	+iPKP	AC	16 07 09.3	Tonga Islands 16.23 S 173.11 W
	ePP	A	10 33	H = 15 47 32.9 h = normal MB=6.1 MS=5.7
	LmH	C	17 05.9	D = 145.46 Az = 354.7 (NEIS)
	LmV	C	05.9	PKPV A 1.2s 762.0nm
				LmH C 22 2.1/um M = 5.8
				LmV C 20 2.6/um 6.0

Day	Phase		h m s	Remarks
5.	ePKP	A	18 26 54.5	Tonga 16.37 S 174.39 W
				H = 18 07 27.8 h = 115 km MB = 4.8
				D = 145.47 Az = 353 (ISC)
5.	ePKP2	A	20 13 38	Tonga 20.3 S 173.5 W
				H = 19 53 40.5 h = 0 km MB = 4.2
				D = 149.44 Az = 352 (ISC)
6.	eP	A	01 14 09.5	Greece - Albania Border Region
	LmH	C	19.8	39.87 N 20.59 E
	LmV	C	19.8	H = 01 11 13.6 h = 63 km MB = 4.1
				D = 12.48 Az = 332.6 (NEIS)
				LmH C 15.5s 0.9 µm
				LmV C 16 1.2 µm
6.	eP	A	01 25 39.5	Hindukush Region 36.44 N 70.13 E
				H = 01 17 55.1 h = 222 km MB = 5.4
				D = 43.60 Az = 308.1 (NEIS)
6.	eP	A	09 56 40	Kamchatka 56.43 N 160.88 E
				H = 09 45 41.8 h = 148 km MB = 4.2
				D = 70.29 Az = 339.8 (NEIS)
6.	eP	A	23 23 50.5	Kurile Islands 44.31 N 147.79 E
	epP	A	24 08	H = 23 11 59.8 h = 89 km MB = 5.3
	esP	A	24 16	D = 78.05 Az = 333.2 (NEIS)
				h = 69 km
				PV A 1.4s 34.9nm M = 5.0
7.	eP	BC	03 55 36	Near Coast of Northern Chile
	ePP	A	59 48.5	26.78 S 70.88 W
	e	BC	59 56	H = 03 41 25.6 h = 25 km MB=5.6 MS=6.1
	LmV	B	04 59.0	D = 105.74 Az = 41.0 (NEIS)
	LmH	B	59.5	LmH B 17.8s 4.4 µm M = 6.1
				LmV B 17 4.5 µm 6.1
7.	ePKIKP	A	06 58 58	South Pacific Cordillera
	LmH	B	08 04.1	54.35 S 136.55 W

Day	Phase		h m s	Remarks
cont. 7.	LmV	B	08 15.3	H = 06 39 00.8 h = normal MB=5.4 MS=6.1
				D = 160.36 Az = 87.8 (NEIS)
				PKIKPV A 1.3s 17.5nm
				LmH B 20 2.5/um M = 5.9
				LmV B 18 3.4/um 6.1
7.	ePKIKP	A	10 22 49	Near Coast of Northern Chile
	LmV	B	11 09.0	26.74 S 70.90 W
	LmH	B	10.0	H = 10 04 26.6 h = 34 km MB=5.7 MS=5.9
				D = 105.71 Az = 41.0 (NEIS)
				LmH B 17s 3.5/um M = 6.0
				LmV B 18 4.6/um 6.1
7.	eP	A	14 37 04	Near Coast of Northern Chile
	e	A	40 31	26.83 S 70.92 W
	e	A	43 20.5	H = 14 22 45.4 h = 14 km
	LmV	B	15 27.4	MB = 5.9 MS = 6.3 (NEIS)
	LmH	B	28.2	D = 105.80
				LmH B 17.5s 11.9/um M = 6.5
				LmV B 18 13.3/um 6.5
7.	eP	A	20 39 11	North Atlantic Ocean 52.48 N 32.38 W
	LmH	B	49.4	H = 20 33 30.0 h = normal MB = 4.5
	LmV	B	49.8	D = 27.10 Az = 76.1 (NEIS)
				LmH B 15.5s 0.6/um M = 4.3
				LmV B 16 0.5/um 4.3
8.	LmH	B	05 38.3	South Sandwich Islands Region
	LmV	B	38.7	58.46 S 24.95 W
				H = 04 37 14 h = 42 km MB = 5.1 (ISC)
				D = 112.7
				LmH B 27.5s 4.6/um M = 5.9
				LmV B 21.5 1.4/um 5.5
8.	eP	A	08 26 18	Albania 41.75 N 19.39 E
	LmH	B	29.8	H = 08 23 47.6 h = 27 km MB = 4.6

Day	Phase		h m s	Remarks
cont. 8.	LmV	B	08 30.7	D = 10.40 Az = 331.5 (NEIS)
				PV A 1.0s 11.8nm M = 5.1
				LmH B 14 0.8 /um 3.7
				LmV B 16 0.8 /um
8.	eP	A	14 38 37.5	Southern Italy 40.82 N 15.39 E
	ePPP	A	38 55.5	H = 14 36 11.0 h = normal MB = 4.7
	LmH	B	42.3	D = 10.17 Az = 346.3 (NEIS)
	LmV	B	43.2	PV A 1.0s 13.8nm M = 5.2
				LmH B 14 4.0 /um 4.4
				LmV B 8 1.7 /um
9.	eP	A	02 30 41	Near Coast of Northern California
	LmH	B	03 06.8	40.26 N 124.23 W
	LmV	B	09.0	H = 02 18 25.8 h = 2 km MB=5.1 MS=4.7
				D = 81.56 Az = 26.6 (NEIS)
				LmH B 20s 0.2 /um M = 4.6
				LmV B 16 0.3 /um 4.8
9.	ePKIKP	A	09 57 15	Solomon Islands 6.09 S 154.46 E
				H = 09 38 59.2 h = 404 km MB = 5.4
				D = 125.86 Az = 331.7 (NEIS)
				PKIKPV A 1.0s 27.6nm
9.	iP	ABC	10 56 22.2	Kurile Islands 43.45 N 146.43 E
	e	A	56 32.5	H = 10 44 26.5 h = 55 km MB = 6.0
	e	A	56 38.5	D = 78.37 Az = 332.5 (NEIS)
	eS	BC	11 06 14	PV A 1.3s 260.0nm M = 6.1
	LmH	B	32.0	PV B 8 1.2 /um 6.1
	LmV	B	37.2	LmH B 17.5 2.9 /um 5.7
				LmV B 18 3.0 /um 5.7
9.	eP	A	11 08 44	Kurile Islands 43.40 N 146.29 E
				H = 10 54 44.4 h = normal MB = 4.8
				D = 78.36 Az = 332.5 (NEIS)

Day	Phase		h m s	Remarks
9.	ePKHKP	ABC	13 26 32	West of Macquarie Island 56.27 S 147.42 E
	e	A	26 51.5	H = 13 06 36.6 h = normal MB=5.6 MS=5.9
	LmH	B	14 38.7	D = 153.46 Az = 276.6 (NEIS)
	LmV	B	52.0	PKHKPV A 1.8s 23.7nm
				LmH B 21 2.3 /um M = 5.9
				LmV B 18 2.2 /um 6.0
9.	ePKIKP	A	19 52 57.5	New Hebrides Islands 19.01 S 168.65 E
				H = 19 33 29.8 h = 45 km MB = 5.0
				D = 143.54 Az = 335.3 (NEIS)
				PKIKPV A 0.9s 11.7nm
9.	eP	A	22 34 48	Off East Coast of Honshu, Japan
				33.84 N 141.69 E
				H = 22 22 13.5 h = 31 km MB = 4.6
				D = 85.05 Az = 330.7 (NEIS)
10.	eP	ABC	00 20 35.5	Off East Coast of Honshu, Japan
	e	A	20 44	34.05 N 141.44 E
	epP	A	20 52.5	H = 00 08 05.8 h = 55 km MB = 5.0
	eS	BC	31 04	D = 84.77 Az = 330.6 (NEIS)
	LmH	B	01 02.2	h = 62 km
	LmV	B	04.8	PV A 2.0s 42.7nm M = 5.2
10.	ePKHKP	A	14 46 56	Tonga Islands 19.15 S 172.98 W
				H = 14 27 06.8 h = 7 km MB = 5.3
				D = 148.37 Az = 354.4 (NEIS)
				PKHKPV A 2.4s 96.7nm
10.	ePKIKP	A	15 03 23	Tonga Islands 19.20 S 172.96 W
	ePKHKP	A	03 26	H = 14 43 41.6 h = normal MB=5.4 MS=5.2
	LmH	B	16 06.0	D = 148.42 Az = 354.4 (NEIS)
	LmV	B	06.0	PKHKPV A 1.8s 44.0nm
10.	eP	A	15 56 47.5	Kurile Islands 43.38 N 146.45 E
				H = 15 44 51.8 h = 59 km MB = 4.8
				D = 78.44 Az = 332.6 (NEIS)

Day	Phase		h m s	Remarks
10.	ePKHKP	A	19 15 06	Tonga Islands 19.17 S 173.18 W
				H = 18 55 20.3 h = normal MB = 4.6
				D = 148.37 Az = 354.2 (NEIS)
11.	ePKHKP	A	04 05 09	Fiji Islands Region 17.80 S 177.00 W
	e	A	06 40.5	H = 03 46 12.3 h = 399 km MB = 5.4
				D = 146.52 Az = 350.0 (NEIS)
				PKHKPV A 1.2s 20.3nm
11.	eP1	ABC	07 26 29	Kansu Province, China 33.00 N 104.02 E
	eP2	A	26 31.2	H = 07 15 39.7 h = normal MB=5.4 MS=6.1
	ePP	A	28 53.5	D = 66.70 Az = 316.2 (NEIS)
	LmH	B	54.7	P2V A 1.7s 127.2nm M = 5.7
	LmV	B	57.2	LmH B 13 15.7 μm 6.4
				LmV B 15.5 10.4 μm 6.2
11.	ePP	A	13 24 20.5	West New Guinea Region 3.46 S 135.61 E
	LmH	B	14 07.7	H = 13 04 35.7 h = 9 km MB = 5.4 (NEIS)
	LmV	B	17.0	D = 113.67
				LmH B 19s 0.5 μm M = 5.1
				LmV B 20 0.4 μm 5.1
11.	eP	A	18 14 21.5	South Atlantic Ridge 28.91 S 13.12 W
	LmV	B	50.0	H = 18 01 59.7 h = normal MB = 4.8
	LmH	B	51.6	D = 82.18 Az = 15.6 (NEIS)
				LmH B 17s 0.25 μm M = 4.8
12.	eP	A	10 29 33.5	Near Coast of Oaxaca, Mexico
				15.74 N 94.62 W
				H = 10 16 45.9 h = 55 km MB = 4.8
				D = 87.91 Az = 37.7 (NEIS)
				PV A 1.4s 14.0nm M = 5.0
12.	eP1	A	15 41 50	Norwegian Sea 72.35 N 1.69 E
	eP2	A	41 55	H = 15 36 55.7 h = normal MB = 4.6
	eS	BC	46 00	D = 22.23 Az = 163.1 (NEIS)
	LmH	B	50.6	P1V A 1.8s 23.6nm M = 4.3

Day	Phase		h m s	Remarks
cont. 12.	LmV	B	15 51.2	P2V A 1.8s 33.8nm M = 4.5
				LmH B 17.5 0.4/um 3.9
				LmV B 18 0.4/um 4.0
13.	eP	A	02 08 53	**West Pakistan** 29.94 N 68.43 E
	LmH	B	33.1	H = 02 00 20.4 h = 7 km MB=4.7 MS=4.3
	LmV	B	35.2	D = 46.73 Az = 313.0 (NEIS)
				PV A 1.5s 15.1nm M = 4.8
				LmH B 14 0.4/um 4.6
				LmV B 12 0.4/um 4.7
13.	ePKIKP	ABC	08 46 47.5	**Near North Coast of New Guinea**
	e	A	47 27	4.47 S 144.04 E
	iPP	BC	48 20	H = 08 28 19.7 h = 112 km MB = 6.0
	ePPP	BC	50 56	D = 119.21 Az = 327.4 (NEIS)
	e	BC	56 00	PKIKPV A 1.8s 43.9nm
	ePKKP	A	57 15	LmH B 19.3 4.3/um
	e	A	57 43	LmV B 17.5 4.2/um
	iSS	BC	09 04 36	
	LmH	B	32.5	
	LmV	B	44.4	
14.	ePP	BC	02 15 28	**Near Coast of Northern Chile**
	ePS	BC	24 50	26.63 S 70.82 W
	LmH	B	54.0	H = 01 57 00.5 h = 29 km MB=5.7 MS=5.5
	LmV	B	54.0	D = 105.59 Az = 40.9 (NEIS)
				LmH B 22.5s 1.2/um M = 5.4
				LmV B 23 1.4/um 5.5
14.	e	A	02 35 12	**Crete** 35.35 N 23.05 E
	e	A	35 51	H = 02 30 45 h = 71 km MB = 4.0 (ISC)
				D = 18.4
14.	ePKIKP	A	04 38 00	**South of Fiji Islands** 22.26 S 179.62 W
	ePKHKP	A	38 06.5	H = 04 19 20.2 h = 592 km MB = 5.2
	ePKP2	A	38 15	D = 150.34 Az = 345.5 (NEIS)
	epPKP	A	40 22.5	PKHKPV A 1.1s 48.4nm
				PKP2V A 1.3 28.4nm

Day	Phase		h m s	Remarks
14.	ePKP	A	05 00 44	<u>Fiji Region</u> 20.8 S 179.3 W
				H = 04 41 01.4 h = 0 km
				D = 148.96 Az = 346 (ISC)
14.	ePg	A	17 44 22.5	<u>Upper Silesia, Poland</u> 50.0 N 18.5 E
	eSg	A	45 22	H = 17 43 00 (BCIS)
				D = 4.45
14.	eP	A	18 32 59.5	<u>West Pakistan</u> 25.44 N 65.66 E
	e	A	33 07	H = 18 24 20.2 h = normal MB = 5.1
	LmH	B	19 01.1	D = 48.12 Az = 316.2 (NEIS)
	LmV	B	01.2	PV A 1.5s 27.6nm M = 5.1
				LmH B 14 0.2 µm 4.3
				LmV B 12 0.3 µm 4.6
15.	eP	A	02 07 21.5	<u>Central Kazakh SSR</u> 42.71 N 67.41 E
				H = 01 59 57.8 h = 0 km MB = 5.3
				D = 38.28 Az = 301.8 (NEIS)
				PV A 1.5s 27.6nm M = 4.7
15.	ePKIKP	A	08 55 00	<u>West Chile Rise</u> 37 S 96 W
				H = 08 36 00 MB = 4.7 (NORSAR)
				D = 128.0
15.	e	A	09 11 49	<u>Burma - India Border Region</u>
				26.42 N 95.47 E
				H = 09 00 54.0 h = 41 km MB = 4.7
				D = 66.28 Az = 316.3 (NEIS)
15.	ePKHKP	A	09 44 50	<u>Easter Island Cordillera</u>
				52.78 S 118.24 W
				H = 09 25 03.7 h = normal MB = 4.8
				D = 149.36 Az = 73.5 (NEIS)
				PKHKPV A 1.4s 46.5nm
15.	ePn	A	14 38 50	<u>Czechoslovakia</u> 49.30 N 13.86 E
				H = 14 38.2 explosion 3.0 t (KHC)
				D = 2.06 Az = 316 (ISC)
				PnV A 1.4s 20.9nm

Day	Phase		h m s	Remarks
15.	eP	A	15 16 58.5	Hindu Kush Region 36.44 N 70.85 E
				H = 15 09 08.8 h = 200 km MB = 4.8
				D = 44.06 Az = 308.1 (NEIS)
16.	eP1	ABC	04 09 35	Yunnan Province, China 23.05 N 101.11 E
	iP2	A	09 38	H = 03 59 10.7 h = normal MB=5.4 MS=6.4
	eS	BC	19 00	D = 72.22 Az = 318.0 (NEIS)
	LmH	C	41.2	P2V A 1.5s 47.7nm M = 5.4
	LmV	C	41.6	LmH C 20 8.8/um 6.0
				LmV C 20 5.1/um 5.8
16.	eP	A	06 16 50.5	Yunnan Province, China 23.19 N 100.94 E
				H = 06 05 28.2 h = normal MB = 5.1
				D = 72.01 Az = 318.0 (NEIS)
16.	eP	ABC	08 12 34	Tibet 33.08 N 86.88 E
	ePP	BC	14 40	H = 08 02 53.8 h = normal MB=5.3 MS=5.5
	eS	BC	20 26	D = 56.30 Az = 312.3 (NEIS)
	LmH	B	37.1	PV A 2.5s 107.7nm M = 5.4
	LmV	B	39.5	LmH B 17 2.6/um 5.4
				LmV B 13.5 1.4/um 5.3
16.	eX	A	11 04 57	Carlsberg Ridge 10.08 N 57.48 E
				H = 10 55 19.5 h = normal MB = 4.9
				D = 55.20 Az = 326.2 (NEIS)
				XV A 1.8s 27.0nm M = 5.0
16.	eP	ABC	12 28 55.5	Andreanof Islands, Aleutian Is.
	eS	BC	38 32	51.29 N 176.64 W
	eSS	BC	44 04	H = 12 16 59.8 h = 47 km MB=5.6 MS=5.8
	LmH	B	13 10.7	D = 78.20 Az = 354.6 (NEIS)
	LmV	B	16.1	PV A 1.5s 37.7nm M = 5.2
				LmH B 17.5 3.5/um 5.7
				LmV B 16 2.3/um 5.7
16.	eP	A	14 37 28	Andreanof Islands, Aleutian Is.
				51.45 N 176.63 W
				H = 14 25 34.4 h = 62 km MB = 5.6
				D = 78.04 Az = 354.6 (NEIS)

Day	Phase		h m s	Remarks
cont. 16.				PV A 1.8s 40.6nm M = 5.1
16.	ePKHKP	A	15 14 18	Fiji Islands Region 21.67 S 179.87 W
				H = 14 55 35.4 h = 620 km MB = 4.7
				D = 149.71 Az = 345.4 (NEIS)
				PKHKPV 0.9s 15.6nm
16.	eSn	A	22 13 40	Yugoslavia 46.0 N 15.7 E
	eSg	A	14 08	H = 22 11 16 (BCIS)
				D = 5.4
16.	eP	A	23 16 26	United Arab Republic 27 N 34 E
				H = 23 10 31 MB = 3.7 (NORSAR)
				D = 29.2
17.	eP	A	01 05 03.5	Kurile Islands 46.41 N 152.68 E
				H = 00 53 11.4 h = 52 km MB = 4.8
				D = 77.64 Az = 335.8 (NEIS)
				PV A 1.0s 11.8nm M = 4.9
17.	eP	ABC	01 54 46.5	Southwestern Ryukyu Islands
	e	A	54 55	23.33 N 123.57 E
	eS	BC	02 05 12	H = 01 42 11.9 h = 28 km MB=5.4 MS=5.4
	LmH	B	36.5	D = 85.13 Az = 323.7 (NEIS)
	LmV	B	37.8	PV A 1.5s 25.1nm M = 5.2
				LmH B 17 3.5 μm 5.8
				LmV B 15 2.9 μm 5.8
17.	eP	A	08 57 27.5	North Atlantic Ridge 45.13 N 28.13 W
	LmH	B	09 07.4	H = 08 51 47.5 h = 33 km MB = 4.5
	LmV	B	08.2	D = 26.98 Az = 64 (ISC)
				PV A 1.7s 12.1nm M = 4.2
				LmH B 14 0.4 μm 4.2
				LmV B 15 0.4 μm 4.2
17.	ePKP	A	16 01 47	New Hebrides Islands 19.88 S 169.62 E
				H = 15 42 41.7 h = 276 km MB = 4.5
				D = 144.71 Az = 335.6 (NEIS)
				PKPV A 0.7s 11.5nm

Day	Phase		h m s	Remarks
17.	eP	A	19 01 29	<u>Panay, Philippine Islands</u>
	LmH	B	49.3	11.50 N 121.50 E
	LmV	B	50.8	H = 18 48 16.8 h = 40 km MB = 5.4
				D = 93.41 Az = 323.1 (NEIS)
				PV A 1.3s 15.3nm M = 5.3
				LmH B 15.5 0.4 µm 5.0
				LmV B 18 0.4 µm 4.9
18.	eP	A	02 20 25.5	<u>Hokkaido, Japan Region</u> 42.01 N 142.48 E
				H = 02 08 32.0 h = 71 km MB = 5.2
				D = 78.24 Az = 330.5 (NEIS)
				PV A 1.0s 21.6nm M = 5.0
18.	eP	A	06 26 11	<u>South Atlantic Ridge</u> 17.65 S 13.54 W
				H = 06 14 50.9 h = normal MB = 4.6
				D = 71.51 Az = 16.6 (NEIS)
18.	eP	AB	08 39 00	<u>Panay, Philippine Islands</u>
	ePP	BC	42 46	11.51 N 121.39 E
	eSKS	BC	49 32	H = 08 25 44.1 h = 14 km MB=5.9 MS=6.1
	eS	C	50 08	D = 93.34 Az = 323.1 (NEIS)
	eSS	C	56 28	PV A 2.0s 81.2nm M = 5.8
	LmH	B	09 23.9	PV B 11 2.1 µm 6.5
	LmV	B	28.5	PPV B 11 2.3 µm 6.5
				LmH B 17 10.7 µm 6.4
				LmV B 17 16.0 µm 6.6
19.	LmH	C	02 50.4	<u>Panay</u> 11.49 N 121.5 E
	LmV	C	51.3	H = 01 45 21.7 h = 22 km (ISC)
				D = 93.5
				LmH C 18s 0.3 µm M = 4.8
				LmV C 16 0.35 µm 4.9
19.	eP	A	19 42 45.5	<u>Mariana Islands Region</u> 21.57 N 142.92 E
	ePP	A	46 41.5	H = 19 29 52.3 h = 326 km MB = 5.4
	ei	A	46 44	D = 96.28 Az = 331.2 (NEIS)
	LmH	C	20 22.7	
	LmV	C	31.3	

Day	Phase		h m s	Remarks
19.	eP	A	23 24 37.5	<u>Greenland Sea</u> 75.06 N 9.87 E H = 23 19 19.8 h = normal MB = 4.4 D = 24.52 Az = 177.3 (NEIS) 1.2s 28.5nm M = 4.7
20.	eP	A	15 21 05.5	<u>Rumania</u> 45.74 N 26.48 E
	LmH	B	25.3	H = 15 18 28.4 h = 73 km MB = 5.6 (NEIS)
	LmV	B	26.3	D = 11.07 PV A 1.4s 41.9nm LmH B 12 1.1/um M = 4.0 LmV B 12 0.9/um
20.	eP	A	22 38 06	<u>Fox Islands, Aleutian Is.</u> 55.8 N 166.9 W H = 22 26 28.5 h = 0 km MB = 4.4 D = 73.93 Az = 1 (ISC)
21.	ePKP	ABC	02 02 48	<u>Fiji Islands Region</u> 18.81 S 176.16 E
	LmV	C	51.6	H = 01 43 11.0 h = 32 km MB = 4.6 (NEIS)
	LmH	C	51.7	D = 145.94 PKPV A 1.6s 22.0nm LmV C 16 0.35/um M = 5.2
21.	eP	A	03 55 25	<u>Fox Islands, Aleutian Is.</u> 53.99 N 166.44 W H = 03 43 48.1 h = 96 km MB = 4.9 (NEIS) D = 75.73 PV A 1.1s 8.1nm M = 4.5
21.	-eP	ABC	14 16 04	<u>Southern Sumatra</u> 5.20 S 102.64 E
	ePS	BC	28 12	H = 14 02 44.8 h = normal
	LmH	B	15 03.4	MB = 5.3 MS = 5.3 (NEIS)
	LmV	B	07.0	D = 94.63 PV A 1.3s 30.6nm M = 5.6 LmH B 21 1.1/um 5.3 LmV B 20 0.7/um 5.2

Day	Phase		h m s	Remarks
21.	ePKP2	A	16 48 26.5	<u>South of Kermadec Islands</u>
				32.22 S 179.97 W
				H = 16 28 17.4 h = 246 km MB = 4.3 (NEIS)
				D = 159.69
21.	e	A	17 29 52	<u>Yugoslavia</u> 45.3 N 16.7 E
	eSn	A	30 40	H = 17 28 01 h = 0 km
	eSg	A	31 22	D = 6.33 Az = 329 (ISC)
22.	eP	A	02 05 46	<u>Sea of Japan</u> 36.82 N 134.61 E
				H = 01 54 19.9 h = 383 km MB = 4.6 (NEIS)
				D = 79.52
22.	ePKIKP	ABC	06 59 16.5	<u>South of Kermadec Islands</u>
	ePKHKP	B	59 30	32.79 S 179.21 W
	ePKP2	B	59 56	H = 06 39 21.4 h = normal MB=5.5 MS=6.0
	ePP	B	07 03 37	D = 160.45 Az = 339.1 (NEIS)
	e	B	09 37	PKIKPV A 2.0s 51.3nm
	e(PSKS)	C	14 04	LmH B 23 4.3 μm M = 6.1
	e	C	17 26	LmV B 20 2.8 μm 6.1
	LmH	B	08 08.3	
	LmV	B	18.0	
22.	eP	ABC	18 25 58.5	<u>Kodiak Islands Region</u> 57.07 N 154.10 W
	ipP	ABC	26 08	H = 18 14 37.2 h = 38 km MB = 5.9 (NEIS)
	eS	BC	35 18	D = 72.01
	LmH	B	56.6	h = 35 km
	LmV	B	57.3	PV A 1.4s 325.0nm M = 6.1
				LmH B 22.5 1.4 μm 5.2
				LmV B 22 1.1 μm 5.1
22.	eP	A	22 25 16.5	<u>Central Mid - Atlantic Ridge</u>
	LmH	B	50.6	0.60 N 25.35 W
	LmV	B	50.9	H = 22 15 17.3 h = normal MB = 5.1
				D = 58.89 Az = 26.6 (NEIS)
22.	eP	A	23 54 27	<u>Oaxaca, Mexico</u> 16.43 N 94.75 W
				H = 23 41 45.9 h = 80 km MB = 4.5
				D = 87.44 Az = 37.7 (NEIS)

Day	Phase		h m s	Remarks
23.	eP	A	14 54 36.5	Rumania 45.66 N 21.12 E
	LmH	B	57.5	H = 14 52 42.3 h = 39 km
	LmV	B	58.8	D = 8.07 Az = 311.6 (NEIS)
				LmH B 13s 1.0/um M = 3.7
				LmV B 9 0.9/um
23.	eP	A	16 58 25	Rumania 45.69 N 21.11 E
	e	A	58 33.5	H = 16 56 27.0 h = 51 km
				D = 8.05 Az = 311.4 (NEIS)
				P traces
23.	ePKIKP	A	17 14 16	New Britain Region 5.38 S 151.49 E
	e	BC	16 00	H = 16 55 25.9 h = 78 km MB = 5.7
	LmH	B	18 05.8	D = 123.81 Az = 330.4 (NEIS)
	LmV	B	11.6	PKIKPV A 1.4s 32.5nm
				LmH B 24 2.0/um
				LmV B 20 1.6/um
23.	e	A	18 26 11	Rumania 46.30 N 21.51 E
				H = 18 23 43.8 h = 86 km
				D = 7.88 Az = 307.1 (NEIS)
23.	eP	A	19 28 49	Honshu, Japan 36.51 N 139.61 E
				H = 19 16 41.0 h = 115 km MB = 5.1 (NEIS)
				D = 81.89
				PV A 1.3s 17.5nm M = 4.8
23.	eP	A	21 10 09	Andreanof Islands, Aleutian Is.
				51.75 N 173.49 W
				H = 20 58 13.7 h = 38 km MB = 4.7
				D = 77.89 Az = 356.7 (NEIS)
				PV A 0.9s 9.7nm M = 4.8
23.	iP	ABC	24 02 52.5	Off East Coast of Honshu, Japan
	eS	BC	13 04	37.21 N 142.13 E
	LmH	B	37.2	H = 23 50 32.4 h = 34 km MB=5.7 MS=5.4
	LmV	B	42.6	D = 82.29 Az = 330.8 (NEIS)
				PV A 1.7s 97.0nm M = 5.6
				LmH B 17 3.0/um 5.7
				LmV B 14 3.8/um 5.7

Day	Phase		h m s	Remarks
24.	eP	A	02 13 23	Southern Iran 27.83 N 52.73 E
	LmH	B	34.2	H = 02 06 01.6 h = normal
	LmV	B	34.3	MB = 5.1 MS = 5.3 (NEIS)
				D = 38.47
				PV A 1.0s 15.7nm M = 4.8
				LmH B 13.5 0.3/um 4.3
				LmV B 14 0.4/um 4.5
24.	eP	A	02 20 03	Southern Iran 28.01 N 52.94 E
				H = 02 12 41.1 h = normal MB = 4.9
				D = 38.46 Az = 317.5 (NEIS)
24.	eP	A	09 41 31	Southern Iran 27.90 N 52.71 E
				H = 09 34 10.5 h = normal MB = 4.9
				D = 38.41 Az = 317.7 (NEIS)
24.	eP	A	10 55 05.5	Strait of Gibraltar 36.20 N 2.04 W
	LmH	B	11 01.8	H = 10 50 59.7 h = normal MB = 3.9
	LmV	B	02.5	D = 17.47 Az = 30.0 (NEIS)
				LmH B 13s 0.9/um M = 4.2
				LmV B 12 1.0/um 4.4
24.	ePKIKP	A	20 37 21	Solomon Islands 7.19 S 156.04 E
	ePP	BC	39 38	H = 20 18 21.4 h = 62 km MB = 5.8 (NEIS)
	LmH	B	21 34.8	D = 127.57
	LmV	B	36.2	PKIKPV A 1.2s 24.4nm
				LmH B 21 0.6/um
				LmV B 18 0.4/um
25.	eP	A	03 58 34.5	West Caroline Islands 10.61 N 138.42 E
	e	A	04 01 25	H = 03 44 34.2 h = normal MB=5.8 MS=5.0
	e	ABC	02 56	D = 103.53 Az = 328.4 (NEIS)
	LmH	B	43.3	LmH B 15.2s 3.8/um M = 6.0
	LmV	B	50.5	LmV B 15 3.1/um 6.0
25.	eP	A	06 23 59	Tonga Islands 17.57 S 174.75 W
				H = 06 04 41.9 h = 200 km MB = 4.4
				D = 146.61 Az = 352.6 (NEIS)

Day	Phase		h m s	Remarks
25.	eP	A	12 27 10.5	Near East Coast of Kamchatka
	LmH	B	13 06.8	56.04 N 163.61 E
	LmV	B	08.7	H = 12 15 53.0 h = normal MB = 5.2
				D = 71.17 Az = 341.6 (NEIS)
				PV A 1.5s 40.2nm M = 5.3
				LmH B 14 0.3 μm 4.7
				LmV B 13 0.3 μm 4.8
25.	eP	A	12 52 40	Crete 34.85 N 24.87 E
				H = 12 48 20.7 h = normal MB = 4.1
				D = 18.50 Az = 332.6 (NEIS)
25.	eP	ABC	14 51 56	Ryukyu Islands Region 25.78 N 129.53 E
	e	A	52 16	H = 14 39 15.1 h = normal MB=5.6 MS=5.4
	ePP	BC	55 20	D = 86.28 Az = 325.7 (NEIS)
	eS	BC	15 02 30	PV A 2.2s 65.4nm M = 5.5
	LmH	B	35.8	LmH B 15.5 2.4 μm 5.7
	LmV	B	35.9	LmV B 14.5 3.0 μm 5.9
25.	eP	A	15 05 48.5	Southern Iran 28.08 N 56.76 E
				H = 14 58 10.8 h = 56 km MB = 5.4 (NEIS)
				D = 40.72
				PV A 1.2s 32.5nm M = 5.0
25.	eP	A	15 27 39.5	Kurile Islands 43.53 N 146.81 E
				H = 15 15 40.8 h = 39 km MB = 5.0
				D = 78.43 Az = 332.7 (NEIS)
				PV A 1.0s 11.8nm M = 4.8
25.	eP	A	16 39 34	Southern Iran 27.98 N 56.71 E
				H = 16 31 57.2 h = 64 km MB = 4.2
				D = 40.76 Az = 316.3 (NEIS)
25.	eP	A	22 02 44	Philippine Islands Region
	LmH	B	39.2	19.50 N 121.57 E
	LmV	B	47.3	H = 21 50 01.4 h = 44 km MB = 5.1
				D = 87.08 Az = 323.2 (NEIS)
				LmH B 16.5s 0.3 μm M = 4.8
				LmV B 13 0.2 μm 4.9

Day	Phase		h m s	Remarks
26.	ePKP	A	01 48 23	New Hebrides Islands 19.73 S 168.11 E
				H = 01 28 49.0 h = 17 km MB = 4.8
				D = 143.98 Az = 334.4 (NEIS)
26.	ePKP	A	18 19 07.5	Fiji Islands Region 17.95 S 178.67 W
	e	A	19 10.5	H = 18 00 37.4 h = 634 km MB = 4.5
				D = 146.36 Az = 348.2 (NEIS)
26.	ePKIKP	ABC	21 47 26	Kermadec Islands 29.57 S 177.45 W
	ePKP2	A	47 58	H = 21 27 34.1 h = 55 km MB = 5.4
				D = 157.87 Az = 344.6 (NEIS)
				PKIKPV A 2.5s 92.2nm
26.	eP	ABC	21 59 07	Andreanof Islands, Aleutian Is.
	e	A	59 37.5	51.25 N 179.26 W
	eS	BC	22 09 16	H = 21 47 12.0 h = 48 km MB=5.2 MS=5.3
	eSS	BC	14 18	D = 78.07 Az = 352.9 (NEIS)
	LmH	B	40.4	PV A 0.8s 19.2nm M = 5.2
	LmV	B	41.7	LmH B 17.5 1.5 /um 5.4
				LmV B 16 1.1 /um 5.3
27.	eP	A	00 19 18.5	Andreanof Islands, Aleutian Is.
				51.46 N 178.39 W
				H = 00 07 26.5 h = 59 km MB = 5.2
				D = 77.92 Az = 353.5 (NEIS)
27.	eP	A	02 23 48	Andreanof Islands, Aleutian Is.
				51.29 N 179.26 W
				H = 02 11 52.9 h = 51 km MB = 4.6
				D = 78.03 Az = 352.9 (NEIS)
27.	eP	A	04 49 00	Andreanof Islands, Aleutian Is.
				51.70 N 173.69 W
				H = 04 37 05.4 h = 45 km MB = 4.8 (NEIS)
				D = 77.94
				PV A 1.0s 9.8nm

Day	Phase		h m s	Remarks
27.	eP	A	06 38 39.5	Andreanof Islands, Aleutian Is. 51.74 N 173.68 W H = 06 26 46.1 h = 51 km MB = 5.1 D = 77.89 Az = 356.6 (NEIS) PV A 1.2s 22.4nm M = 5.1
27.	eP	A	06 57 38.5	Andreanof Islands, Aleutian Is. 51.11 N 179.04 W H = 06 45 44.4 h = 62 km MB = 4.5 D = 78.22 Az = 353.1 (NEIS)
27.	e(P)	A	11 27 30	Czechoslovakia 48.85 N 14.35 E Explosion 7.3 t H = 11 26.5 (PRU) D = 2.53 Az = 316 (ISC)
27.	LmH	B	13 09.5	Off Coast of Southern Chile 45.2 S 77.1 W H = 12 01 22 h = 42 km MB = 4.7 (ISC) D = 122.3 LmH B 22s 0.5/um LmV B 20 0.4/um
	LmV	B	11.0	
27.	ePKIKP	BC	14 07 57	New Hebrides Islands 15.99 S 168.10 E H = 13 48 31.6 h = 11 km MB = 5.7 MS = 5.6 (NEIS) D = 140.59 LmH B 19s 0.7/um M = 5.4 LmV B 18.5 0.7/um 5.5
	ePP	BC	11 02	
	LmH	B	15 26.2	
	LmV	B	28.7	
27.	eP	A	14 15 45	Andreanof Islands, Aleutian Is. 51.2 N 179.3 W H = 14 03 51.2 h = 60 km MB = 4.5 (NEIS) D = 78.15
27.	ePKP	A	14 41 41.5	Fiji Islands Region 18.10 S 178.42 W H = 14 23 06.1 h = 610 km MB = 5.1 D = 146.55 Az = 348.4 (NEIS) PKPV A 1.2s 24.4nm

Day	Phase		h m s	Remarks
27.	ePKP	A	14 45 36	Fiji Islands Region 18.07 S 178.54 W
				H = 14 27 02.2 h = 631 km MB = 5.0 (NEIS)
				D = 146.50
				PKPV A 1.6s 27.5nm
27.	eP	A	16 51 27	Andreanof Islands, Aleutian Is.
				51.71 N 173.77 W
				H = 16 39 33.7 h = 58 km MB = 4.5
				D = 77.92 Az = 356.5 (NEIS)
28.	eP	A	03 06 56	Central Kazakh SSR 50.55 N 68.40 E
	e	A	07 51.5	H = 02 59 57.6 h = 0 km MB = 5.3
				D = 35.28 Az = 292.8 (NEIS
				Underground explosion (UPP)
				PV A 0.8s 15.4nm M = 4.9
28.	-iP	AB	10 03 18.5	Vera Cruz, Mexico 18.27 N 96.60 W
	-eipP	AB	03 37	H = 09 50 40.0 h = 84 km MB = 6.8
	eisP	A	03 47	D = 87.07 Az = 37.3 (NEIS)
	-iPP	BC	06 48	h = 70 km
	eiPPP	B	09 07	PV A 2.2s 1790.0nm M = 6.8
	ei	BC	13 06	PH A 2.0 980.0nm
	iSKS	BC	13 40	pPV A 2.1 1360.0nm
	ePS	B	15 20	sPV A 2.2 3840.0nm
	iSS	C	19 30	PPV B 12 11.5/um 7.2
	ePKKP	A	21 06.5	SH B 13 6.7/um 6.5
	LmH	B	43.3	LmH B 20.5 41.8/um
	LmV	B	43.5	LmV B 20 47.3/um
28.	eP	A	13 40 41	Near E. Coast of Honshu, Japan
	e	A	40 49	36.57 N 141.34 E
				H = 13 28 19.0 h = 25 km MB = 4.2 (NEIS)
				D = 82.54
				P traces
28.	+eP1	AB	15 11 40.5	Central Mid-Atlantic Ridge
	iP2	A	11 43.5	0.19 S 18.03 W
	eiP3	A	11 47.5	H = 15 01 59.1 h = normal

Day	Phase		h m s	Remarks
cont. 28.	eiP4	A	15 11 58	MB = 5.8 MS = 6.8 (NEIS)
	ePcP	B	12 55	D = 56.58
	ePP	B	14 00	P1V A 1.0s 35.4nm M = 5.4
	iPPP	B	15 17	P1V B 7 1.1/um 6.0
	iS	B	19 43	P2V A 1.9 220.0nm 5.9
	LmH	B	39.4	P3V A 2.2 687.0nm 6.3
	LmV	B	39.4	P4V A 2.8 3480.0nm 6.9
				PPV B 10 11.2/um 7.0
				PPPV B 11 15.5/um
				SH B 13.5 60.3/um 7.3
				LmH B 15 122.8/um 7.1
				LmV B 13.5 154.5/um 7.3
28.	e	A	21 02 29	Region of Parma, Italy 44.9 N 10.5 E
	eSn	A	03 47.5	H = 21 00 21 (BCIS)
				D = 5.3
29.	ePKP	A	08 09 45	Tonga Islands 17.09 S 173.23 W
	e	A	10 03.5	H = 07 50 05.0 h = normal MB = 5.2 (NEIS)
				D = 146.30
				PKPV A 1.6s 19.2nm
29.	e(PKP2)	A	17 04 26.5	South of Fiji Islands 26 S 179 E
				H = 16 44 28 MB = 3.7 (NORSAR)
				D = 153.6
				traces
29.	eP	A	19 24 56.5	Ionian Sea 37.92 N 20.06 E
				H = 19 21 34.4 h = 0 km MB = 3.7
				D = 14.07 Az = 337 (ISC)
30.	eP	A	01 26 09	Kurile Islands 44 N 147 E
				H = 01 14 13 MB = 4.3 (NORSAR)
				D = 78.0
				PV A 1.0s 9.8nm
30.	e(P)	A	07 41 52.5	Turkey 38.03 N 42.70 E
	e	A	43 36.5	H = 07 36 23.6 h = normal MB=4.8 MS=4.7

Day	Phase		h m s	Remarks
cont. 30.	eS	BC	07 46 26	D = 25.35 Az = 309.9 (NEIS)
	LmH	B	52.9	LmH B 17.5s 3.6 µm M = 5.0
	LmV	B	54.3	LmV B 13.5 2.0 µm 4.9
30.	ePKIKP	ABC	09 14 32	<u>Off E. Coast of N. Island, N.Z.</u>
	e	A	15 35.5	37.31 S 179.45 E
	LmH	B	10 31.1	H = 08 54 32.4 h = 46 km MB=5.8 MS=5.4
	LmV	B	33.5	D = 164.08 Az = 330.7 (NEIS)
				LmH B 21s 1.3 µm M = 5.6
				LmV B 19 1.4 µm 5.8
30.	ePKIKP	A	13 19 36	<u>South of Fiji Islands</u> 24.26 S 179.96 E
	ePKHKP	A	19 43.5	H = 13 00 45.9 h = 505 km MB = 4.9
				D = 152.15 Az = 344.0 (NEIS)
				PKIKPV A 1.6s 16.5nm
30.	ePg	A	15 05 20	<u>Hilders/Rhön, Fed. Rep. of Germany</u>
				50°32.54' N 10°02.27' E
				H = 15 05 00.64 yield 8.625 t (Hannover)
				D = 1.0
30.	eP	ABC	18 37 37.5	<u>Northern Colombia</u> 7.35 N 72.83 W
	epP	ABC	38 22.5	H = 18 25 43.1 h = 181 MB = 5.7 (NEIS)
	e	BC	38 40	D = 80.85 h = 180 km
	esPP	BC	41 37	PV A 1.6s 35.8nm M = 4.8
	eS	BC	47 32	PV B 12 1.4 µm 5.5
	esS	BC	48 48	LmH B 17 1.4 µm
	LmV	B	19 14.3	LmV B 16 1.2 µm
	LmH	B	14.4	
30.	eP	ABC	20 01 35	<u>Bay of Bengal</u> 7.10 N 84.31 E
	ipP	A	01 48	H = 19 50 02.9 h = normal MB=5.9 MS=5.2
	eS	BC	11 04	D = 73.58 Az = 320.7 (NEIS)
	LmH	B	37.0	h = 50 km
	LmV	B	40.3	PV A 2.0s 350.0nm M = 6.0
				LmH B 17.5 1.4 µm 5.3
				LmV B 16 0.9 µm 5.2

Day	Phase		h m s	Remarks
30.	eP	A	21 12 56	West Pakistan 30.17 N 68.11 E
				H = 21 04 30.3 h = normal MB = 4.7
				D = 46.37 Az = 312.8 (NEIS)
				PV A 1.1s 8.1nm M = 4.6
30.	eP	A	21 26 29	South of Honshu, Japan 32.48 N 141.65 E
				H = 21 13 48.3 h = 39 km MB = 4.9 (NEIS)
				D = 86.22
				PV A 1.5s 12.6nm
30.	ePKIKP	A	23 27 17	Fiji Islands Region 20.96 S 178.85 W
	ePKHKP	A	27 22.5	H = 23 08 41.7 h = 627 km MB = 5.1
	ePKP2	A	27 29.5	D = 149.24 Az = 346.9 (NEIS)
				PKHKPV A 1.5s 50.2nm
				PKP2V A 1.2 24.4nm
31.	eP	A	02 41 45	Southern Alaska 61.10 N 147.41 W
	e	A	41 49.5	H = 02 30 57.9 h = 49 km MB=5.1 MS=5.0
	epP	A	41 58	D = 67.35 Az = 14.3 (NEIS)
	LmH	B	03 15.3	h = 50 km
	LmV	B	15.4	LmH B 18s 0.35 μm M = 4.6
				LmV B 16 0.35 μm 4.7
31.	eP	A	05 02 31	Eastern Caucasus 43.47 N 45.47 E
				H = 04 57 17.0 h = normal MB=4.6 MS=3.4
				D = 24.00 Az = 299.3 (NEIS)
				PV A 1.9s 49.2nm M = 4.7
31.	ePKP	A	23 49 25	Samoa Islands Region 16.09 S 172.49 W
				H = 23 29 49.4 h = normal MB = 5.1
				D = 145.37 Az = 355.4 (NEIS)

Day	Phase		h m s	Remarks
1.	ePKP2	A	05 58 49	<u>Kermadec Islands Region</u> 28.20 S 177.05 W H = 05 38 30.4 h = 57 km MB = 4.4 D = 156.63 Az = 346.0 (NEIS)
1.	eP	A	11 34 30	<u>Zambia</u> 13.31 S 24.12 E H = 11 23 54.6 h = normal MB = 4.7 D = 64.61 Az = 351.2 (NEIS) traces
1.	LmH LmV	B B	13 00.4 00.4	LmH B 11s 0.25/um LmV B 11 0.45/um
1.	e(PKP2)	A	18 27 02	<u>South of Fiji Islands</u> 22.81 S 176.27 W H = 18 07 08.8 h = normal MB = 5.2 D = 151.54 Az = 349.4 (NEIS)
2.	eP ePP	A A	00 41 58.5 46 06	<u>Mindanao, Philippine Islands</u> 7.41 N 123.74 E H = 00 29 26.2 h = 619 km M = 5.3 (NEIS) D = 97.99 PV A 1.5s 20.1nm M = 5.2
2.	eP	A	04 06 56	<u>Dead Sea Region</u> 32.80 N 35.3 E H = 04 01 29 h = 0 km D = 24.94 Az = 323 (ISC)
2.	eP e e LmH LmV	A A A B B	07 31 49 31 57 32 05 58.7 08 00.6	<u>Near Coast of West Pakistan</u> 24.90 N 63.14 E H = 07 23 17.1 h = 25 km MB = 5.3 D = 46.95 Az = 317.0 (NEIS) PV A 1.6s 33.0nm M = 5.1 LmH B 18 0.7/um 4.6 LmV B 13 0.8/um 4.9
2.	ePKP	A	16 26 03.5	<u>New Hebrides Islands Region</u> 19.25 S 167.62 E H = 16 06 33.5 h = normal MB = 5.2 D = 143.35 Az = 334.3 (NEIS) PV A 1.6s 16.5nm

Day	Phase		h m s	Remarks				
3.	ePKIKP	A	03 42 21.5	**Fiji Islands Region** 20.89 S 178.96 W				
	-iPKHKP	AB	42 26.5	H = 03 23 47.4 h = 639 km MB = 5.4				
	+iPKP2	A	42 33.5	D = 149.16 Az = 346.8 (NEIS)				
				PKHKPV	A	1.3s	91.7nm	
				PKP2V	A	1.4	55.8nm	
4.	ePKP2	A	14 48 26	**Kermadec Islands** 30.16 S 177.61 E				
				H = 14 27 53.7 h = 16 km MB = 4.5 (NEIS)				
				D = 158.40				
4.	eP	AB	17 37 48	**Near Coast of Oaxaca, Mexico**				
	ePP	BC	41 12	14.98 N 94.30 W				
	eS	BC	48 24	H = 17 24 59.4 h = 51 km MB = 5.2				
	LmH	B	18 16.8	D = 88.32 Az = 37.8 (NEIS)				
	LmV	B	22.3	PV	A	2.0s	77.0nm	M = 5.6
				PV	B	12	0.9 μm	5.9
				LmH	B	18	2.0 μm	5.6
				LmV	B	16.5	2.6 μm	5.8
4.	ePKIKP	AB	21 01 50	**Kermadec Islands** 30.37 S 177.90 W				
	ePKP2	A	02 26	H = 20 41 55.4 h = normal MB=5.4 MS=5.3				
	ePP	A	06 03	D = 158.52 Az = 343.3 (NEIS)				
				LmH	B	23s	0.6 μm	M = 5.3
				LmV	B	20	0.4 μm	5.3
5.	-iP	AB	00 42 34	**Near West Coast of Honshu, Japan**				
	epP	A	43 17	40.66 N 139.55 E				
	eS	BC	52 10	H = 00 30 53.5 h = 191 km MB = 5.6				
	LmH	B	01 15.3	D = 78.31 Az = 329.2 (NEIS)				
				h = 160 km				
				PV	A	1.6s	77.0nm	M = 5.2
				LmH	B	17	0.4 μm	
5.	eP	A	01 19 32	**Andreanof Islands, Aleutian Is.**				
				51.70 N 173.42 W				
				H = 01 07 36.4 h = 42 km MB = 4.9				
				D = 77.94 Az = 356.7 (NEIS)				
				PV	A	1.2s	14.2nm	M = 4.9

Day	Phase		h m s	Remarks
5.	ePKP	A	02 48 46	Samoa Islands Region 14.49 S 175.88 W
				H = 02 29 14.9 h = normal MB = 4.4
				D = 143.43 Az = 352.0 (NEIS)
				traces
5.	ePKIKP	A	04 03 17	Fiji Islands Region 19.68 S 177.90 W
	ePKHKP	A	03 22	H = 03 44 21.0 h = 402 km MB = 5.0
	ePKP2	A	03 26.5	D = 148.20 Az = 348.5 (NEIS)
	epPKP	A	05 04	
5.	ePKP	A	05 48 27	Fiji Region 19.4 S 178.4 W
				H = 05 28 46 h = 0 km MB = 4.8
				D = 147.84 Az = 348 (ISC)
5.	eP	A	09 20 12	Kurile Islands 43.95 N 148.33 E
				H = 09 08 14.0 h = 47 km MB = 4.6
				D = 78.54 Az = 333.6 (NEIS)
				PV A 1.1s 12.1nm M = 4.8
5.	+eP	AB	13 15 24.5	Off East Coast of Honshu, Japan
	ePP	B	18 29	39.49 N 143.11 E
	e	A	18 35	H = 13 03 13.9 h = 41 km MB=5.5 MS=5.8
	ePPP	B	20 18	D = 80.68 Az = 331.1 (NEIS)
	eS	B	25 31	PV A 1.7s 78.8nm M = 5.4
	eSS	C	30 48	PV B 16 1.6 µm 5.7
	LmH	B	54.3	PPV B 13 1.0 µm 5.9
	LmV	B	56.4	SH B 17 2.4 µm 5.9
				LmH B 15 14.3 µm 6.4
				LmV B 15 12.9 µm 6.4
5.	eP	A	14 21 12	Off East Coast of Honshu, Japan
				39.47 N 143.18 E
				H = 14 09 00.2 h = normal MB = 4.7
				D = 80.72 Az = 331.1 (NEIS)
				PV A 1.9s 15.2nm M = 4.7

Day	Phase		h m s	Remarks
5.	ePKHKP	A	15 03 23	Tonga Islands 20.89 S 174.09 W
	e	A	03 32	H = 14 43 37.1 h = 65 km MB = 4.4
				D = 149.97 Az = 352.7 (NEIS)
5.	eP	A	16 15 51	Off East Coast of Honshu, Japan
				39.47 N 143.30 E
				H = 16 03 37.2 h = 18 km MB = 4.5
				D = 80.76 Az = 331.2 (NEIS)
6.	eP	A	01 18 12	Off W. Coast of Northern Sumatra
				4.40 N 94.92 E
				H = 01 05 50.3 h = 41 km MB = 4.8
				D = 82.37 Az = 320.4 (NEIS)
6.	eP	A	01 36 27	Northern Sumatra 0.34 N 98.06 E
				H = 01 23 40.5 h = normal MB = 4.7
				D = 87.47 Az = 320.5 (NEIS)
6.	eP	AB	11 10 30.5	Southern Alaska 61.04 N 146.83 W
	eipP	A	10 39	H = 10 59 36.7 h = 29 km MB=5.5 MS=5.3
	eS	B	19 28	D = 67.33 Az = 14.7 (NEIS)
	LmV	B	45.7	h = 35 km
	LmH	B	46.3	PV A 1.6s 44.0nm M = 5.3
				LmH B 12 0.4 μm 4.8
				LmV B 14 0.4 μm 4.9
6.	eSg	A	13 24 16.5	Poland 50.35 N 18.90 E
	e	A	24 23	H = 13 21 43.2 M = 2.8 (WAR)
				D = 4.6
7.	ePn	A	00 12(09)	Rockburst in the mining district of
	ePg	A	12 15	the Ruhr, Federal Republic of Germany
	eSn	A	12(38)	H = 00 11 25 (BCIS)
	eSg	A	12 47	D = 2.61 Az = 113 (ISC)
7.	e	A	01 04 05	

Day	Phase		h m s	Remarks
7.	ePKIKP	A	09 50 07.5	<u>New Britain Region</u> 5.62 S 151.50 E
	LmH	B	10 47.1	H = 09 31 14.8 h = 61 km MB = 5.5
	LmV	B	47.3	D = 124.03 Az = 330.3 (NEIS)
				PKIKPV A 1.2s 10.2nm
				LmH B 20 0.4 μm
				LmV B 20 0.4 μm
7.	eP	A	11 22 09.5	<u>North of Svalbard</u> 81.90 N 1.39 W
	LmH	B	12 27.1	H = 11 15 47.0 h = normal MB=4.6 MS=3.7
	LmV	B	27.2	D = 31.65 Az = 164.2 (NEIS)
				PV A 1.0s 15.7nm M = 4.8
				LmH B 16 0.6 μm 4.4
				LmV B 16 0.6 μm 4.5
7.	ePKIKP	A	14 17 25	<u>New Britain Region</u> 5.55 S 151.11 E
	ePP	A	19 22	H = 13 58 33.6 h = 71 km MB = 5.6
	LmH	B	15 09.8	D = 123.77 Az = 330.2 (NEIS)
	LmV	B	11.4	LmH B 24s 1.4 μm
				LmV B 22 1.3 μm
7.	eP	A	18 59 25	<u>Off East Coast of Honshu, Japan</u>
	e	A	59 33	39.90 N 143.49 E
				H = 18 47 14.5 h = 33 km MB = 4.1
				D = 80.46 Az = 331.3 (NEIS)
7.	eP	A	19 40 27.5	<u>Rumania</u> 45.80 N 26.48 E
				H = 19 37 51.8 h = 140 km MB = 4.6 (NEIS)
				D = 11.04
				PV A 1.4s 34.9nm M = 4.8
8.	eP	A	01 25 46.5	<u>Andreanof Islands, Aleutian Is.</u>
				51.30 N 179.23 W
				H = 01 13 52.2 h = 54 km MB = 4.9
				D = 78.02 Az = 353.0 (NEIS)
				PV A 0.9s 17.5nm M = 5.0

Day	Phase		h m s	Remarks
8.	eP	A	02 05 44	<u>Turkey</u> 36.13 N 31.23 E
				H = 02 01 12.0 h = 75 km
				D = 20.25 Az = 322 (ISC)
				PV A 1.4s 11.6nm M = 4.0
8.	e	A	04 36 02.5	<u>Near S. Coast of Honshu, Japan</u>
				34.48 N 139.40 E
				H = 04 23 26.4 h = 9 km MB = 4.7 (NEIS)
				D = 83.55
8.	eP	AB	07 35 22	<u>Tibet</u> 33.24 N 86.75 E
	ePP	AB	37 29.5	H = 07 25 43.9 h = normal
	ePPP	B	38 40	MB = 5.5 MS = 5.2 (NEIS)
	eS	B	43 12	D = 56.11
	eSS	C	47 00	PV A 1.9s 87.1nm M = 5.5
	LmH	B	59.9	PV B 11 1.2 μm 5.9
	LmV	B	08 01.2	PPV B 9 0.6 μm 5.7
				LmH B 17 10.8 μm 6.0
				LmV B 14.5 5.6 μm 5.9
8.	eP	A	07 59 52	<u>Peru-Brazil Border Region</u>
				8.3 S 74.2 W
				H = 07 46 52.69 h = 168 km MB = 5.6 (NEIS)
				D = 93.72
				PV A 1.6s 60.5nm M = 5.6
8.	eP1	A	08 09 09.5	<u>Tibet</u> 33.18 N 86.72 E
	eP2	A	09 11	H = 07 59 31.2 h = normal MB = 5.1
	ePP	A	11 16.5	D = 56.14 Az = 312.2 (NEIS)
				P1V A traces
				P2V A 1.4s 23.2nm M = 5.1
8.	LmV	B	16 57.0	<u>El Salvador</u> 13.2 N 89.8 W
	LmH	B	57.4	H = 16 00 03 h = 62 km MB = 4.2 (ISC)
				D = 87.0
				LmV B 17s 0.3 μm

Day	Phase		h m s	Remarks
8.	ePKIKP	A	20 01 10	<u>Kermadec Islands</u> 29.21 S 178.54 W
	ePKHKP	A	01 22	H = 19 41 32.02 h = 142 km MB = 4.9 (NEIS)
	ePKP2	A	01 43	D = 157.26
9.	-eP	AB	02 24 20	<u>Szechwan Province, China</u>
	ePcP	A	24 52.5	31.65 N 100.01 E
	eS	B	33 06	H = 02 13 39.4 h = normal MB = 5.5
	LmV	B	55.9	D = 65.30 Az = 315.5 (NEIS)
	LmH	B	56.0	PV A 1.9s 60.6nm M = 5.4
				PV B 5 0.45/um 5.8
				LmH B 12.5 0.8/um 5.1
				LmV B 12 0.8/um 5.1
9.	eP	A	02 53 15.5	<u>Szechwan Province, China</u>
				31.48 N 100.00 E
				H = 02 42 33.1 h = normal MB = 5.2
				D = 65.41 Az = 315.6 (NEIS)
9.	eP	A	05 11 16	<u>Off East Coast of Honshu, Japan</u>
				39.93 N 143.43 E
				H = 04 59 04.7 h = 25 km MB = 4.0
				D = 80.41 Az = 331.2 (NEIS)
				PV A traces
9.	eP	A	05 28 43.5	<u>Off East Coast of Honshu, Japan</u>
	epP	A	28 51.5	39.89 N 143.54 E
				H = 05 16 31.7 h = 23 km MB = 4.6
				D = 80.49 Az = 331.3 (NEIS)
				h = 29 km
				PV A 0.9s 9.7nm M = 4.8
9.	eP	AB	08 42 28	<u>Ascension Island Region</u>
	e	A	42 33.5	7.11 S 12.84 W
	e	A	42 41	H = 08 32 14.8 h = normal MB=5.2 MS=4.9
	eS	B	50 49	D = 61.30 Az = 17.5 (NEIS)
	eSS	B	54 28	PV A 1.4s 34.9nm M = 5.3
	LmH	B	09 07.5	SH B 13 0.6/um 5.4
	LmV	B	10.3	LmH B 18 0.7/um 4.8
				LmV B 18 1.0/um 5.1

Day	Phase		h m s	Remarks
9.	eP1	AB	18 38 02	Off East Coast of Honshu, Japan
	iP2	A	38 03.5	39.54 N 143.14 E
	+ePP	B	41 10	H = 18 25 49.4 h = 23 km MB=5.7 MS=5.9
	eS	B	48 08	D = 80.65 Az = 331.1 (NEIS)
	LmV	B	19 19.2	P1V A traces
	LmH	B	19.4	P2V A 1.4s 74.4nm M = 5.5
				P1V B 14 1.9/um 6.0
				PPV B 13 1.5/um 6.1
				SH B 15 2.7/um 6.1
				LmH B 15.5 15.7/um 6.5
				LmV B 16 15.5/um 6.5
9.	eP	A	20 21 24	Off East Coast of Honshu, Japan
	epP	A	21 35	39.39 N 143.25 E
				H = 20 09 12.0 h = 28 km MB = 5.0
				D = 80.82 Az = 331.2 (NEIS)
				h = 37 km
				PV A 1.2s 16.3nm M = 4.9
9.	PKIKP	A	20 36 36	Easter Islands Cordillera
				34.99 S 109.54 W
				H = 20 17 16.39 h = normal MB = 5.2
				D = 135.29 Az = 50.8 (NEIS)
9.	eP	A	21 05 14	Mid - Indian Rise 14.49 S 66.04 E
				H = 20 53 04.1 h = normal
				D = 80.38 Az = 328.3 (NEIS)
10.	eP	AB	03 07 10	Turkey 38.49 N 39.62 E
	eS	C	11 25	H = 03 02 04.2 h = normal MB = 4.9
	eSPP	BC	11 36	D = 23.21 Az = 310.7 (NEIS)
	LmH	B	18.8	LmH B 13.5s 0.3/um M = 4.0
	LmV	B	18.9	LmV B 14 0.3/um 4.2
10.	eP	A	04 53 24	Shikoku, Japan 33.68 N 132.65 E
	epP	A	53 35	H = 04 41 11.5 h = 55 km MB = 4.3
				D = 81.26 Az = 326.5 (NEIS)
				h = 39 km

Day	Phase		h m s	Remarks
10.	eP	A	07 04 56.5	Kodiak Island Region 57.32 N 154.10 W
	e	A	05 16.5	H = 06 53 34.7 h = 28 km MB = 4.7
				D = 71.76 Az = 9.5 (NEIS)
				PV A 1.0s 13.8nm M = 5.0
10.	-eiP	AB	07 54 08.5	E. Russia - N.E. China Border Region
	-ipP	AB	56 05	42.45 N 130.91 E
	epPP	B	58 45	H = 07 43 30.5 h = 532 km MB = 6.0
	-iS	BE	08 02 51	D = 73.26 Az = 324.6 (NEIS)
	+iS	BN	02 51	PV A 1.7s 940.0nm M = 6.0
	+iScS	BN	03 19	PV B 10 3.8 µm 5.9
	esS	B	06 16	SH B 12 16.1 µm 6.6
	e(P'P')	A	21 34	LmH B 12 6.2 µm
	e(SKPP')	A	24 24	LmV B 13 6.6 µm
	LmV	B	30.1	
	LmH	B	30.3	
10.	eP	A	17 08 30	Off East Coast of Honshu, Japan
	LmH	B	44.0	39.45 N 143.47 E
	LmV	B	49.8	H = 16 56 17.7 h = 30 km MB = 4.6
				D = 80.84 Az = 331.3 (NEIS)
				LmH B 16s 0.3 µm M = 4.7
				LmV B 16 0.2 µm 4.7
10.	ePg	A	21 13 19.5	D c. 1.0
	eSg	A	13 33	
11.	eP	A	10 30 09	Chagos Archipelago Region
				3.42 S 68.13 E
				H = 10 18 46.2 h = normal MB = 5.5
				D = 72.24 Az = 326.1 (NEIS)
11.	LmH	B	19 48.0	Off East Coast of Honshu, Japan
	LmV	B	48.0	39.65 N 143.32 E
				H = 18 54 49 h = 29 km MB = 4.3 (ISC)
				D = 80.6
				LmH B 15.5s 0.4 µm M = 4.9
				LmV B 15 0.4 µm 4.9

Day	Phase		h m s	Remarks
11.	-eP	AB	23 31 05	Northeast of Taiwan 25.65 N 124.51 E
	e	A	31 20	H = 23 18 50.8 h = 141 km MB = 5.8
	epP	B	31 40	D = 83.78 Az = 323.8 (NEIS)
	ePP	B	34 20	h = 138 km
	epPP	B	35 06	PV A 3.0s 1000.0nm M = 6.1
	eS	B	41 13	PH A 3.0 518.0nm 6.1
	esS	B	42 15	PV B 10 2.6 /um 6.0
	LmH	B	24 10.2	PPV B 13 1.0 /um 5.8
	LmV	B	18.2	SH B 13 1.9 /um 6.1
				LmH B 16.5 5.2 /um
				LmV B 15.5 3.2 /um
12.	eP	A	01 30 49.5	Dodecanese Islands 36.53 N 26.92 E
				H = 01 26 48.6 h = 152 km MB = 4.2
				D = 17.89 Az = 326.8 (NEIS)
				PV A 1.5s 20.1nm M = 4.2
12.	+iP	AB	07 06 00	Novaya Zemlya 73.30 N 55.16 E
	eiPn	B	06 27	H = 06 59 54.3 h = 0 km MB=6.8 MS=5.0
	eiPcP	B	09 07	D = 29.35 Az = 243.5 (NEIS)
	eiS	B	10 54	Underground explosion MB = 7.1 (UPP)
	eiSn	B	11 31	PV A 1.0s 1200.0nm M = 6.7
	LmH	B	17.9	PV B 4 3.2 /um 6.5
	LmV	B	20.9	LmH B 12 47.1 /um 6.3
				LmV B 8 27.9 /um 6.4
12.	eP	A	09 39 38	Greece 40.74 N 20.99 E
				H = 09 36 50.5 h = 93 km MB = 4.6
				D = 11.86 Az = 329.7 (NEIS)
12.	e	A	19 00 59	
12.	eP	A	20 24 02	Kurile Islands 45.95 N 149.47 E
				H = 20 12 25.1 h = 164 km MB = 4.8
				D = 77.11 Az = 334.0 (NEIS)
				PV A 1.4s 18.6nm M = 4.6

Day	Phase		h m s	Remarks
12.	ePKP	A	23 15 14	South of Fiji Islands 26 S 179 W
				H = 22 55 19 (NORSAR)
				D = 154.2
				PKPV A 1.3s 13.1nm
13.	+eP	A	06 33 15.5	Mindanao, Philippine Islands
	epP	A	33 31	9.16 N 126.14 E
	LmV	B	07 22.2	H = 06 19 45.8 h = 68 km MB = 5.8
	LmH	B	22.3	D = 97.99 Az = 324.2 (NEIS)
				h = 59 km
				PV A 1.3s 28.4nm M = 5.6
14.	LmH	B	11 56.6	Afghanistan-USSR Border Region
	LmV	B	56.8	38.29 N 70.4 E
				H = 11 29 53.9 h = 86 km MB = 4.5 (ISC)
				D = 42.5
14.	ePKHKP	A	12 58 08.5	Fiji Islands Region 17.92 S 178.73 W
	ePKP2	A	58 11	H = 12 39 31.3 h = 580 km MB = 4.4
				D = 146.32 Az = 348.1 (NEIS)
14.	ePg	A	15 05 30	Dorheim/Hessen, Fed. Republic of Germany
				50°57.88' N 09°13.15' E
				Explosion
				H = 15 05 00.83 yield 3.1 t (Hannover)
				D = 1.5
14.	ePKP2	A	23 57 06.5	South of Fiji 26.2 S 179.4 W
				H = 23 36 55.8 h = 0 km
				D = 154.14 Az = 344 (ISC)
				PV A 1.8s 27.0nm
15.	+eP	AB	01 50 57.5	Iceland Region 63.90 N 22.20 W
	Pm	A	02 00 02.5	H = 01 45 57.7 h = 1 km MB=5.3 MS=5.4
	eS	B	55 07	D = 22.25 Az = 110.6 (NEIS)
	LmV	B	02.6	PV A 1.1s 36.3nm M = 4.8
	LmH	B	02.7	PmV A 1.1 171.4nm 5.4
				PV B 11 1.0/um 5.2

Day	Phase		h m s	Remarks
cont. 15.	LmH	B	02 02.7	SH B 10.5s 3.3/um M = 5.7 LmH B 12 7.5/um 5.4 LmV B 11.5 9.2/um 5.6
15.	eP	A	02 27 10	Iceland Region 63.87 N 22.14 W H = 02 22 15.7 h = normal MB = 4.9 D = 22.21 Az = 110.6 (NEIS) PV A 1.7s 51.5nm M = 4.7
15.	eP	A	04 37 36	Ryukyu Islands 29.34 N 130.26 E
	eS	B	48 00	H = 04 25 08.7 h = 34 km MB = 5.3
	LmV	B	05 19.0	D = 83.72 Az = 325.8 (NEIS)
	LmH	B	19.1	PV A 1.4s 41.8nm M = 5.4 LmH B 17 1.3/um 5.4 LmV B 17 1.7/um 5.5
15.	ePKP	A	12 56 30	Tonga 15.9 S 174.1 W H = 12 36 50 h = 0 km D = 145.06 Az = 354 (ISC)
15.	eP	A	16 10 55	Hokkaido Region 43.07 N 143.79 E H = 15 58 10.8 h = 131 km MB = 4.5 D = 77.79 Az = 331 (ISC)
15.	LmH	B	16 14.7	Caroline Islands Region
	LmV	B	15.3	7.54 N 144.60 E H = 15 05 25.3 h = 12 km MB = 5.0 (ISC) D = 109.4 LmV B 19s 0.35/um M = 4.9
15.	ePKP	A	17 19 10	Tonga Islands 15.44 S 174.91 W H = 16 59 35.2 h = normal MB = 4.5 D = 144.49 Az = 352.8 (NEIS) PKPV A 1.6s 16.5nm
15.	ePKHKP	A	17 52 44	Tonga Islands Region 22.54 S 175.40 W H = 17 32 55.4 h = 39 km MB = 4.9 D = 151.41 Az = 350.6 (NEIS) PKHKPV A 1.3s 8.7nm

Day	Phase		h m s	Remarks
15.	ePKIKP	A	23 28 05	Solomon Islands 6.83 S 155.33 E
				H = 23 09 06.4 h = 63 km MB = 4.9
				D = 126.92 Az = 331.9 (NEIS)
				PKIKPV A traces
16.	LmV	B	03 24.8	LmH B 12s 0.3/um
	LmH	B	24.9	LmV B 12 0.4/um
16.	eP	AB	05 05 29.5	South of Panama 5.23 N 78.03 W
	eS	B	15 56	H = 04 52 50.9 h = normal MB=5.4 MS=5.6
	LmH	B	38.3	D = 85.77 Az = 39.7 (NEIS)
	LmV	B	50.5	PV A 2.0s 59.8nm M = 5.5
				LmH B 22 1.0/um 5.2
				LmV B 16 0.6/um 5.1
16.	eP	AB	08 34 15.5	South of Panama 5.24 N 78.11 W
	eS	B	44 44	H = 08 21 38.1 h = normal MB=5.3 MS=5.2
	LmV	B	09 06.7	D = 85.82 Az = 39.7 (NEIS)
	LmH	B	07.2	PV A 1.6s 27.5nm M = 5.2
				PV B 5.5 0.4/um 5.2
				LmH B 20 0.4/um 4.8
				LmV B 24 0.8/um 5.1
16.	e	A	19 22 35	Kyushu, Japan 30.16 N 131.11 E
	LmH	B	20 04.7	H = 19 09 59.1 h = 30 km MB = 4.7
	LmV	B	04.9	D = 83.45 Az = 326.1 (NEIS)
				LmH B 15s 0.5/um M = 5.0
				LmV B 14 0.7/um 5.2
16.	eP1	AB	21 31 53.5	Iceland Region 63.94 N 22.23 N
	eiP2	AB	31 56.3	H = 21 26 53.5 h = 2 km MB=5.2 MS=5.1
	eS	BC	36 05	D = 22.27 Az = 110.7 (NEIS)
	LmH	B	43.6	P1V A 1.8s 74.4nm M = 4.9
	LmV	B	43.6	P2V A 1.3 146.0nm 5.3
				SH B 11 2.4/um 5.4
				LmH B 12 7.9/um 5.4
				LmV B 12 10.0/um 5.7

Day	Phase		h m s	Remarks
16.	eX	A	21 59 02	XV A 1.5s 10.1nm
16.	eP	A	22 38 27	Iceland Region 64.87 N 22.35 W
	e	A	38 58	H = 22 33 28.4 h = 5 km MB = 4.7
				D = 22.30 Az = 110.3 (NEIS)
16.	ePKP	A	23 11 27.5	Fiji 17.3 S 179.7 E
				H = 22 51 46 h = 0 km
				D = 145.42 Az = 347 (ISC)
17.	eP	A	01 19 10	Iceland 63.99 N 22.22 W
				H = 01 14 17.2 h = normal MB = 4.1
				D = 22.29 Az = 110.8 (NEIS)
				traces
17.	eP	A	04 12 22	Iran 36.52 N 51.11 E
	e	A	14 18.5	H = 04 06 03.7 h = 47 km MB = 4.8
	LmV	B	30.0	D = 31.49 Az = 309.2 (NEIS)
	LmH	B	30.7	LmH B 15s 0.25/um M = 4.0
				LmV B 15 0.45/um 4.2
17.	ePKHKP	A	06 39 55	Tonga Region 23.2 S 174.6 W
				H = 06 19 59.9 h = 33 km
				D = 152.17 Az = 352 (ISC)
17.	ePKHKP	A	07 15 35.5	South of Australia 50.52 S 139.67 E
	LmH	B	54.3	H = 06 55 53.4 h = normal
	LmV	B	54.3	D = 147.57 Az = 290.8 (NEIS)
				LmH B 17s 0.3/um M = 5.1
				LmV B 18 0.4/um 5.3
17.	ePKP	AB	07 41 13	Tonga Islands 17.32 S 174.34 W
	epPKP	A	41 48	H = 07 21 47.4 h = 135 km MB = 5.5
				D = 146.41 Az = 353.1 (NEIS)
				h = 131 km
				PKPV A 1.5s 211.0nm

Day	Phase		h m s	Remarks
17.	-eP	A	14 51 03	Greece 38.54 N 20.09 E
	LmH	B	57.5	H = 14 47 47.4 h = 4 km MB = 4.6
	LmV	B	57.5	D = 13.52 Az = 336.3 (NEIS)
				PV A 1.0s 15.8nm M = 5.0
				LmH B 10 0.6 µm 5.4
				LmV B 8 0.55 µm 5.4
17.	ePP	A	15 41 25	Mediterranean Sea 35.85 N 21.86 E
				H = 15 37 20.4 h = 43 km MB = 4.3
				D = 16.54 Az = 336.6 (NEIS)
				PPV A 0.7s 9.6nm
17.	ePKHKP	A	15 51 32	Tonga Islands Region 23.83 S 175.24 W
	ePKP2	A	51 44.5	H = 15 31 37.2 h = normal MB = 4.6
				D = 152.71 Az = 350.5 (NEIS)
				PKHKPV A 1.2s 12.2nm
17.	e(P)	A	23 12 52.5	Off Coast of Oregon 44.59 N 129.26 W
				H = 23 00 37.3 h = 14 km MB = 4.4 (NEIS)
				D = 79.24
17.	eP1	A	23 45 38	Off Coast of Oregon 44.43 N 129.31 W
	eP2	A	45 41	H = 23 33 33.0 h = normal MB=5.1 MS=4.6
	LmV	B	24 23.0	D = 79.41 Az = 24.1 (NEIS)
	LmH	B	25.0	LmH B 14s 0.4 µm M = 4.9
				LmV B 16 0.4 µm 4.8
18.	e(P)	A	00 41 47	Off Coast of Oregon 44.49 N 129.41 W
				H = 00 29 36.9 h = normal MB=4.8 MS=4.3
				D = 79.38 Az = 24.0 (NEIS)
18.	e	A	03 03 43	
18.	+eP	A	08 52 07.5	Turkey 36.68 N 30.14 E
	ei	A	52 17	H = 08 47 40.5 h = 15 km MB = 4.4
	LmV	B	09 00.6	D = 19.28 Az = 322.2 (NEIS)
	LmH	B	00.7	PV A 1.8s 40.6nm M = 4.4
				LmH B 11 0.8 µm 4.3
				LmV B 12 0.6 µm 4.3

Day	Phase		h m s	Remarks
18.	eP	AB	13 11 29	Northern Peru 7.03 S 76.14 W
				H = 12 58 25.2 h = 133 km MB = 5.8
				D = 93.96 Az = 39.6 (NEIS)
				PV A 2.0s 111.0nm M = 5.8
18.	ePKIKP	AB	13 52 48	South Pacific Cordillera
	ePKHKP	A	52 58	54.52 S 132.62 W
	e	A	53 08.5	H = 13 32 51.6 h = normal MB=5.3 MS=6.4
	eSKKS	BC	14 03 50	D = 158.08 Az = 85.4 (NEIS)
	eSKSP	C	07 28	PKIKPV A 2.6s 69.4nm
	ePPS	C	10 24	LmH B 18.5 4.0/um M = 6.2
	eSS	BC	17 08	LmV B 15 3.0/um 6.2
	LmV	B	15 08.6	
	LmH	B	09.3	
19.	eP	A	03 07 11	Central Kazakh SSR 45.64 N 67.85 E
	ePP	A	08 35	H = 02 59 57.2 h = 0 km MB = 5.2
	ePcP	A	09 43.5	D = 37.09 Az = 298.6 (NEIS)
				Underground explosion (UPP)
				PV A 1.0s 27.6nm M = 4.9
19.	eP	A	03 56 13.5	
19.	eP	A	10 43 38	Kodiak Island Region 56.23 N 154.25 W
				H = 10 32 09.6 h = normal MB=5.0 MS=4.2
				D = 72.85 Az = 9.4 (NEIS)
				PV A 0.9s 13.6nm M = 4.9
19.	eP	A	11 13 16	South of Honshu, Japan 33.20 N 140.81 E
				H = 11 00 44.9 h = 60 km MB = 5.0
				D = 85.24 Az = 330.3 (NEIS)
				PV A 1.6s 16.5nm M = 4.9
19.	eSn	A	23 33 45	Pyrenees 43.26 N 0.30 W
	eSg	A	34 57	H = 23 28 58.7 h = 0 km (ISC)
				D = 11.0

Day	Phase		h m s	Remarks
20.	ePKP	A	14 17 22	Fiji Islands 15.61 S 179.95 E
	LmH	B	15 24.9	H = 13 57 45.0 h = 11 km MB=5.3 MS=4.8
	LmV	B	25.1	D = 143.81 Az = 347.4 (NEIS)
20.	eP	A	19 09 58	Andreanof Islands, Aleutian Is.
				51.63 N 173.87 W
				H = 18 58 01.4 h = 39 km MB = 4.5
				D = 78.00 Az = 356.4 (NEIS)
20.	eP1	A	20 56 10.5	Negros, Philippine Islands
	-eiP2	AB	56 12.5	9.05 N 123.79 E
	epP	AB	58 17	H = 20 43 39.8 h = 560 km MB = 6.0
	eiPP	AB	21 00 15	D = 96.71 Az = 323.6 (NEIS)
	epPP	B	02 26	h = 580 km
	eSKS	B	05 55	P1V A 1.3s 35.0nm M = 5.5
	eS	B	06 44	P2V A 1.5 231.0nm 6.3
	ePS	B	09 25	P2V B 8 1.52 μm 6.4
	eSS	B	13 41	SH B 10 3.4 μm 6.1
	esSS	C	16 48	LmH B 17 3.4 μm
	eSSS	C	17 28	LmV B 14 2.4 μm
	LmH	B	38.5	
	LmV	B	49.9	
21.	ePn	A	01 40 03	Yugoslavia 44.25 N 17.10 E
	e	A	41 42	H = 01 38 12.2 h = normal
	e	A	41 59	D = 7.40 Az = 331.8 (NEIS)
	eSg	A	42 05	
	e	A	42 12	
21.	eP	A	02 33 18.5	Southern Honshu, Japan 35.10 N 134.43 E
	e	A	33 22.5	H = 02 21 04.1 h = 23 km MB = 5.1
				D = 80.88 Az = 327.2 (NEIS)
21.	eP	A	05 18 58	Near S. Coast of Honshu, Japan
	e	A	19 19	33.31 N 137.66 E
	e	A	19 38.5	H = 05 07 05.9 h = 352 km MB = 4.7
	epP	A	20 19	D = 83.82 Az = 328.8 (NEIS)
				h = 357 km

Day	Phase		h m s	Remarks
21.	e(PKIKP)	AB	07 32 12	Northern Easter I. Cordillera
				4.39 S 102.00 W
				H = 07 13 34.0 h = normal MB=6.1 MS=6.0
				D = 108.24 Az = 37.9 (NEIS)
21.	ePKIKP	A	07 49 29	Northern Easter I. Cordillera
	e	A	49 51	4.36 S 101.94 W
				H = 07 31 02.8 h = normal MB=6.0 MS=5.6
				D = 108.18 Az = 37.9 (NEIS)
				PV A 1.8s 27.1nm
21.	eP	A	09 40 41	Off East Coast Kamchatka 53.30 N 161.60 E
				H = 09 29 10.3 h = normal MB = 4.0
				D = 73.39 Az = 340.6 (NEIS)
				traces
21.	e	A	08 00 42	traces
21.	eP	AB	10 47 15	Kodiak Island Region 57.25 N 154.01 W
	e	AB	47 26	H = 10 35 55.9 h = 45 km MB = 5.0
	e	A	47 35	D = 71.82 Az = 9.6 (NEIS)
				PV A 1.4s 51.2nm M = 5.3
21.	LmH	B	14 45.0	Luzon 18.65 N 120.77 E
	LmV	B	47.0	H = 13 48 37.0 h = 66 km MB = 5.0 (ISC)
				D = 87.3
21.	ePKIKP	A	19 47 08.5	South of Fiji Islands 26.12 S 178.28 E
	ePKHKP	AB	47 17.5	H = 19 28 29.4 h = 651 km MB = 5.6
	ePKP2	AB	47 32	D = 153.46 Az = 340.8 (NEIS)
	e(PP)	B	51 07	PKIKPV A 1.8s 47.3nm
	ePPP	B	54 39	PKHKPV A 1.5 75.4nm
				PKP2V A 1.3 214.0nm
22.	eP	A	02 36 25	North Atlantic Ridge 26.58 N 44.67 W
				H = 02 27 42.3 h = normal MB=4.7 MS=5.9
				D = 48.73 Az = 44.8 (NEIS)

Day	Phase		h m s	Remarks
22.	eP	A	02 36 25	North Atlantic Ridge 26.58 N 44.67 W
				H = 02 27 42.3 h = normal MB=4.7 MS=5.9
				D = 48.73 Az = 44.8 (NEIS)
22.	eP	A	03 06 03	North Atlantic Ridge 26.53 N 44.55 W
	e	AB	06 07	H = 02 57 19.5 h = normal MB=5.1 MS=5.6
	eiX	A	06 30.5	D = 48.69 Az = 44.8 (NEIS)
	eiY	A	06 35	PV A 1.0s 11.8nm M = 4.9
	eiS	B	13 09	XV A 1.8 67.6nm
	LmH	B	22	YV A 1.6 66.0nm
	LmV	B	22.2	SH B 15 2.1/um 5.8
				LmH B 24 5.2/um 5.4
				LmV B 24 4.9/um 5.5
22.	eP	A	05 15 16	Philippine Islands Region
	LmV	B	06 10.9	9.96 N 126.31 E
	LmH	B	11.0	H = 05 01 44.2 h = 36 km MB=5.3 MS=4.7
				D = 97.44 Az = 324.3 (NEIS)
				PV A 2.0s 42.7nm M = 5.6
				LmH B 16 0.6/um 5.2
				LmV B 16 0.6/um 5.2
22.	eP	A	06 33 31.5	Southern Greece 36.54 N 23.97 E
	e	A	39 26.5	H = 06 29 42.0 h = 92 km MB = 4.5
				D = 16.66 Az = 331.6 (NEIS)
22.	eP	A	18 48 20.5	Andreanof Islands, Aleutian Is.
	LmH	B	19 31.0	51.64 N 173.69 W
	LmV	B	31.0	H = 18 36 24.6 h = 36 km MB = 5.0
				D = 77.99 Az = 356.5 (NEIS)
				PV A 1.1s 28.2nm M = 5.2
22.	+iP	A	19 39 44.5	Andreanof Islands, Aleutian Is.
	epP	A	39 57	51.64 N 173.69 W
	LmH	B	20 26.0	H = 19 27 49.3 h = 49 km MB=5.4 MS=4.5
	LmV	B	26.0	D = 77.99 Az = 356.6 (NEIS)
				h = 46 km
				PV A 1.1s 44.4nm M = 5.4

Day	Phase		h m s	Remarks
23.	eP	A	02 30 21	<u>Near East Coast of Kamchatka</u>
				54.72 N 161.77 E
				H = 02 19 07.3 h = 110 km MB = 4.5
				D = 72.08 Az = 340.5 (NEIS)
23.	eP	A	07 38 08	<u>Kurile Islands</u> 43.77 N 145.97 E
				H = 07 26 11.5 h = 39 km MB = 4.9
				D = 77.93 Az = 332.3 (NEIS)
23.	eP	A	12 39 18.5	<u>Leyte, Philippine Islands</u>
	ePP	A	43 08.5	10.35 N 125.29 E
	LmH	B	13 19.8	H = 12 25 52.1 h = 39 km MB=5.7 MS=5.1
	LmV	B	24.5	D = 96.54 Az = 324.1 (NEIS)
				PV A 1.2s 16.3nm M = 5.4
				LmH B 19 0.8 µm 5.2
				LmV B 18 0.8 µm 5.3
23.	e	A	14 09 53.5	
23.	eP	A	16 54 06	<u>Mindanao, Philippine Islands</u>
	LmH	B	17 43.3	9.94 N 126.35 E
	LmV	B	43.5	H = 16 40 34.0 h = 46 km MB=5.2 MS=4.9
				D = 97.48 Az = 324.3 (NEIS)
				PV A 1.6s 16.5nm M = 5.3
				LmH B 16 0.4 µm 5.0
				LmV B 16 0.5 µm 5.1
23.	ePKHKP	A	18 01 49	<u>South of Fiji Islands</u> 22.11 S 179.73 W
	ePKP2	A	01 56	H = 17 43 04.3 h = 609 km MB = 4.6
				D = 150.16 Az = 345.4 (NEIS)
24.	ePKIKP	A	03 11 09.5	<u>Fiji Islands Region</u> 20.08 S 178.42 W
	ePKHKP	A	11 13.5	H = 02 52 31.5 h = 576 km MB = 4.3
				D = 148.48 Az = 347.7 (NEIS)
				PKHKPV A 1.4s 16.3nm

Day	Phase		h m s	Remarks
24.	eP	A	09 15 38	North of Svalbard 86.13 N 32.03 E H = 09 08 39.4 h = normal MB=4.9 MS=4.9 D = 35.92 Az = 202.2 (NEIS) PV A 1.0s 27.6nm M = 5.1
24.	eP	A	13 53 17.5	Kurile Islands 43.49 N 146.58 E
	epP	A	53 28	H = 13 41 20.4 h = 46 km MB = 4.9 D = 78.39 Az = 332.6 (NEIS) h = 39 km PV A 1.0s 11.8nm
24.	eP	A	22 37 20	North of Svalbard 86.08 N 22.0 E H = 22 30 19.3 h = 0 km MB = 4.5 D = 35.67 Az = 191 (ISC) PV A 1.5s 12.6nm M = 4.6
25.	ePKHKP	A	02 26 26	Tonga Islands 20.52 S 174.88 W H = 02 06 38.9 h = normal MB = 4.9 D = 149.50 Az = 351.8 (NEIS)
25.	eP	A	05 34 07.5	Eastern Mediterranean Sea
	e	A	34 15.5	34.85 N 28.35 E H = 05 29 40.9 h = 79 km MB = 4.1 D = 19.93 Az = 327.5 (NEIS) PV A 1.3s 8.7nm M = 3.9
25.	e	A	10 13 26.5	
25.	eP	A	13 08 00.7	Hindu Kush Region 36.46 N 70.85 E
	e	A	13 09 16	H = 13 00 17.4 h = 197 km MB = 4.8 D = 44.04 Az = 308 (ISC)
25.	ePKIKP	A	16 37 17	West of Macquarie Island
	ePKHKP	AB	37 25	54.85 S 145.85 E
	ePKP2	A	37 35.5	H = 16 17 28.3 h = normal MB=5.9 MS=6.3
	LmH	B	17 54.5	D = 152.37 Az = 280.4 (NEIS)
	LmV	B	56.5	PKIKPV A 1.9s 30.3nm PKHKPV A 2.2 240.0nm LmH B 19 6.0/um M = 6.3 LmV B 19 5.9/um 6.4

Day	Phase		h m s	Remarks
25.	eP	A	21 08 51	Off East Coast of Honshu, Japan 39.38 N 143.47 E H = 20 56 38.3 h = 31 km MB = 4.1 D = 80.90 Az = 331.3 (NEIS) traces
25.	LmH	B	23 32.0	Either West Irian Region 1.19 S 136.66 E
	LmV	B	32.8	H = 22 20 49 h = 10 km (ISC) D = 112.5 or Mindanao 7.59 N 126.91 E H = 22 26 30.9 h = 84 km MB = 5.4 (ISC) D = 99.8 LmH B 18s 0.25/um LmV B 17 0.3/um
26.	e(PKP)	A	00 24 53	Fiji Islands Region 19.39 S 177.66 W H = 00 05 53.1 h = 411 km MB = 4.5 D = 147.96 Az = 348.8 (NEIS) traces
26.	eP	A	10 00 17.5	Leyte, Philippine Islands
	LmH	B	39.4	10.22 N 125.27 E
	LmV	B	48.3	H = 09 46 48.8 h = normal MB=5.4 MS=5.3 D = 96.64 Az = 324.0 (NEIS) PV A 1.7s 24.2nm M = 5.4 LmH B 22 2.2/um 5.6 LmV B 19 2.0/um 5.6
26.	ePKHKP	A	16 47 40	West of Macquarie Island
	ePKP2	A	47 46.5	55.36 S 146.25 E H = 16 27 47.2 h = normal MB=5.5 MS=4.9 D = 152.69 Az = 279.2 (NEIS) PKPV A 2.0s 29.9nm
26.	LmV	B	18 21.7	LmH B 18s 0.25/um
	LmH	B	22.0	LmV B 18 0.4/um

Day	Phase		h m s	Remarks
26.	+ePKP	A	19 42 46.5	<u>Tonga</u> 16.79 S 106.15 W
				H = 19 23 07.3 h = 25 km MB = 4.5
				D = 145.98 Az = 354 (ISC)
				PV A 1.0s 21.6 nm
27.	eP	A	02 12 49	<u>West Pakistan</u> 33.85 N 72.15 E
				H = 02 04 23.0 h = 35 km MB = 4.9
				D = 46.51 Az = 310.2 (NEIS)
27.	+iP	AB	07 05 50.6	<u>Novaya Zemlya</u> 70.76 N 53.87 E
	LmH	B	17.7	H = 06 59 58.0 h = 0 km MB=6.0 MS=4.9
	LmV	B	17.7	D = 27.90 Az = 246.2 (NEIS)
				Underground explosion MB = 5.9 (UPP)
				PV A 1.2s 130.0nm M = 5.6
				LmH B 11 0.7 µm 4.5
				LmV B 12 0.5 µm 4.5
27.	eP1	AB	12 34 08.5	<u>Jan Mayen Island Region</u>
	eP2	A	34 12	71.54 N 12.10 W
	eS	BC	38 28	H = 12 29 04.3 h = normal MB=5.1 MS=5.6
	LmH	B	44.1	D = 23.56 Az = 140.2 (NEIS)
	LmV	B	48.9	P2V A 2.0s 120.0nm M = 5.3
				SH B 12 2.7 µm 5.6
				LmH B 16.7 4.8 µm 5.1
				LmV B 11 2.5 µm 5.1
27.	eP	A	22 16 00.5	<u>Eastern Gulf of Aden</u> 13.22 N 50.68 E
				H = 22 07 16.7 h = normal MB = 5.0
				D = 48.95 Az = 327.9 (NEIS)
				PV A 1.6s 38.2nm
28.	eP	A	01 11 32	<u>Northern Sumatra</u> 1.91 N 99.15 E
	epP	A	12 02	H = 00 58 59.5 h = 132 km MB = 5.0
				D = 86.97 Az = 320.4 (NEIS)
				h = 117 km
				PV A 1.6s 22.0nm M = 4.9

Day	Phase		h m s	Remarks
28.	-eP	A	11 41 08	<u>Eastern Gulf of Aden</u> 13.24 N 50.72 E
				H = 11 32 23.3 h = normal MB = 5.5
				D = 48.96 Az = 327.8 (NEIS)
				PV A 1.5s 90.5nm M = 5.6
28.	eP	A	22 44 02	<u>Eastern Gulf of Aden</u> 13.3 N 50.6 E
				H = 22 35 18 h = 33 km MB = 4.4
				D = 48.89 Az = 328 (ISC)
28.	ePKHKP	A	24 11 40	<u>West of Macquarie Island</u>
				55.35 S 146.00 E
				H = 23 51 46.9 h = normal MB = 5.1
				D = 152.54 Az = 279.3 (NEIS)
29.	-iP	AB	00 54 37.7	<u>North Korea</u> 41.89 N 130.87 E
	epP	B	56 29	H = 00 44 00.78 h = 575 km MB = 6.5
	+ipP	B	56 44	D = 73.70 Az = 324.6 (NEIS)
	+isP	B	57 42	PV A off scale
	eiS	B	01 03 22	PV B 7s 26.4 μm M = 6.9
	eisS	B	07 00	SH B 17 146.3 μm 7.5
	ei	B	08 05	LmH B 14 80.2 μm
	ei	B	12 52	LmV B 14 75.2 μm
	e	A	21 18.5	
	e(SKPPKP)	A	24 57	
	LmH	B	28.9	
	LmV	B	30.7	
29.	ePKP2	A	03 24 43.5	<u>West of Macquarie Island</u>
				60.26 S 150.4 E
				H = 03 04 23.0
				D = 155.11 Az = 266 (ISC)
				PKP2V A 1.0s 13.8nm
30.	eP	A	05 05 40	<u>Western Russia</u> 51.61 N 54.58 E
				H = 04 59 57.5 h = 0 km MB = 5.2
				D = 26.70 Az = 285.1 (NEIS)
				Underground explosion MB = 5.9 (UPP)
				PV A 1.2s 32.5nm M = 4.9

Day	Phase		h m s	Remarks
30.	+iP	AB	06 30 13	<u>Near East Coast of Honshu, Japan</u>
	LmH	B	07 09.8	35.61 N 140.45 E
	LmV	B	12.6	H = 06 17 52.8 h = 62 km MB = 5.9
				D = 83.01 Az = 330.0 (NEIS)
				PV A 1.4s 153.0nm M = 5.8
				LmH B 15.5 2.7/um
				LmV B 14 2.5/um
30.	ePcP	A	20 37 50	<u>Near Coast of Venezuela</u> 10.55 N 62.36 W
	LmH	B	21 07.0	H = 20 26 13 h = 0 km MB = 4.8
	LmV	B	08.8	D = 71.75 Az = 40 (ISC)

Day	Phase		h m s	Remarks
1.	LmH	B	06 04.0	Mindanao 6.36 N 124.14 E
	LmV	B	04.3	H = 04 56 33 h = 60 km MB = 5.1 (ISC)
				D = 99.0
				LmH B 18s 0.6 /um
				LmV B 17 0.7 /um
1.	+eP	AB	14 28 43	Near East Coast of Honshu, Japan
	epP	A	28 56.5	35.72 N 140.56 E
	ePP	B	31 55	H = 14 16 23.0 h = 56 km MB = 5.6
	ePP	B	32 00	D = 82.96 Az = 330.1 (NEIS)
	eS	B	39 00	h = 48 km
	eSS	B	45 14	PV A 1.6s 60.4nm M = 5.4
	LmH	B	15 08.4	PV B 6 0.71 /um 5.8
	LmV	B	11.1	PPV A 2.0 76.9nm 5.7
				LmH B 17 2.3 /um
				LmV B 13.5 2.5 /um
1.	LmH	B	20 33.5	Near Coast of Oaxaca, Mexico
	LmV	B	39.5	15.26 N 94.47 W
				H = 19 41 53.3 h = 39 km MB = 4.9 (ISC)
				D = 88.2
				LmH B 18s 0.25 /um M = 4.7
				LmV B 17 0.35 /um 4.9
1.	ePKIKP	A	24 02 43.5	New Britain Region 4.54 S 151.47 E
				H = 23 44 12.0 h = 226 km MB = 5.4
				D = 123.08 Az = 330.7 (NEIS)
				PKIKPV A 1.2s 14.2nm
2.	eP	A	00 21 49	Mongolia 44.43 N 102.62 E
				H = 00 11 57.1 h = normal MB = 4.7
				D = 57.99 Az = 311.4 (NEIS)
2.	eP	A	03 12 07	Taiwan 23.90 N 121.63 E
	epP	A	12 19	H = 02 59 42.3 h = 49 km MB=5.1 MS=4.8
	LmH	B	54.4	D = 83.61 Az = 323.0 (NEIS)
	LmV	B	54.4	h = 43 km
				LmH B 16s 1.2 /um M = 5.4
				LmV B 16 1.4 /um 5.5

Day	Phase		h m s	Remarks
2.	ePKP	A	07 20 02.5	Tonga 18.6 S 175.40 W
	LmH	B	08 28.0	H = 07 00 18.3 h = 33 km MB = 4.5
	LmV	B	34.6	D = 147.57 Az = 352 (ISC)
				LmH B 20s 0.25/um M = 4.9
				LmV B 18 0.4/um 5.2
2.	eP	A	08 41 33	Hokkaido, Japan Region 42.89 N 145.30 E
				H = 08 29 35.8 h = 52 km MB = 5.0
				D = 78.48 Az = 332.0 (NEIS)
				PV A 1.0s 15.8nm M = 4.9
2.	ePP	A	12 41 40	Bali Island Region 8.75 S 115.36 E
				H = 12 23 23.3 h = 92 km MB = 5.6
				D = 105.43 Az = 320.1 (NEIS)
3.	eP	A	04 40 31	Baja California 27.56 N 112.42 W
	LmV	B	05 22.7	H = 04 27 40.5 h = 9 km MB=5.5 MS=4.7
	LmH	B	22.9	D = 87.79 Az = 31.9 (NEIS)
				PV A 1.7s 36.4nm M = 5.4
				LmH B 15 0.8/um 5.3
				LmV B 14 0.7/um 5.3
3.	ePKIKP	A	08 58 11	South of Fiji Islands 22.21 S 176.36 W
	ePKHKP	A	58 17	H = 08 38 33.7 h = 91 km MB = 4.8
	e	A	58 48	D = 150.94 Az = 349.5 (NEIS)
				PKHKPV A 1.5s 25.1nm
3.	eP	A	08 59 50.5	Andreanof Islands, Aleutian Is.
				50.12 N 179.13 W
				H = 08 47 47.6 h = 37 km MB = 4.8 (NEIS)
3.	eP	A	09 29 33.5	Western Caucasus 43.28 N 41.04 E
				H = 09 24 23.5 h = 0 km MB = 4.1
				D = 21.29 Az = 301 (ISC)
				PV A 1.4s 20.9nm M = 4.4
3.	eP	A	10 43 22	Black Sea 42.8 N 40.78 E
				H = 10 38 35.4 h = 33 km MB = 3.3
				D = 21.35 Az = 302 (ISC)

Day	Phase		h m s	Remarks
3.	eP	AB	10 47 48.5	Kurile Islands 45.54 N 151.78 E
	LmH	B	11 21.8	H = 10 35 51.3 h = normal MB=5.2 MS=4.8
	LmV	B	27.0	D = 78.17 Az = 335.4 (NEIS)
				LmH B 17.5s 1.3 µm M = 5.3
				LmV B 16 0.8 µm 5.2
3.	e	A	11 12 25	
3.	ePKIKP	A	11 28 20	South of Kermadec Islands
	ePKHKP	A	28 34	32.63 S 179.13 W
	ePKP2	A	29 02	H = 11 08 27.2 h = 37 km MB=5.4 MS=5.3
	eX	A	29 14.5	D = 160.33 Az = 339.3 (NEIS)
	LmH	B	12 37.5	PKIKPV A 2.2s 65.5nm
	LmV	B	53.9	LmH B 22.5 1.4 µm M = 5.6
				LmV B 18 0.6 µm 5.5
3.	eP	A	12 53 49.5	Off East Coast of Honshu, Japan
				37.75 N 142.05 E
				H = 12 41 34.1 h = 50 km MB = 4.7
				D = 81.79 Az = 330.7 (NEIS)
				traces
3.	eP	A	13 06 49	Kurile Islands 45.74 N 151.75 E
				H = 12 54 52.3 h = normal MB = 4.7
				D = 77.98 Az = 335.3 (NEIS)
				PV A 1.2s 8.1nm M = 4.6
3.	ePKP	A	16 10 35	Fiji Islands Region 17.66 S 176.94 W
	e	A	11 12	H = 15 51 35.0 h = 360 km MB = 4.9
				D = 146.39 Az = 350.1 (NEIS)
				PKPV A 1.3s 26.2nm
4.	eP	A	13 50 41.5	Kurile Islands 45.75 N 151.71 E
				H = 13 38 44.6 h = normal MB = 4.8
				D = 77.96 Az = 335.3 (NEIS)
				PV A 1.4s 27.9nm M = 5.1

Day	Phase		h m s	Remarks
4.	ePKIKP	A	18 18 22.5	Fiji Islands Region 21.12 S 179.20 W
	ePKHKP	A	18 27.5	H = 17 59 48.3 h = 638 km MB = 5.0
	ePKP2	A	18 35	D = 149.33 Az = 346.5 (NEIS)
				PKHKPV A 1.2s 36.6nm
				PKP2V A 1.2 24.4nm
5.	e(PP)	A	04 56 04	Southern Greece 36.34 N 21.21 E
				H = 04 52 11.1 h = 27 km MB = 4.1
				D = 15.88 Az = 337.2 (NEIS)
5.	eP dif	B	06 00 05	Near Coast of Central Chile
	e	B	02 22	32.99 S 71.92 W
	ePKIKP	A	04 00	H = 05 45 27.3 h = 14 km MB = 5.8 (NEIS)
	ePP	AB	04 35	D = 110.95
	ePPP	B	06 52	PPV A 3.8s 520.0nm M = 6.5
	eSKS	B	10 40	PPV B 20 1.9 µm 6.4
	eiPS	B	14 15	LmH B 17 38.6 µm 7.1
	LmH	B	55.4	LmV B 17.5 53.0 µm 7.2
	LmV	B	55.4	
5.	ePn	A	10 28 16.5	Yugoslavia 45.54 N 14.31 E
	eSn	A	29 15.5	H = 10 26 55.2 h = 18 km
	eSg	A	29 47	D = 5.42 Az = 341.5 (NEIS)
5.	iPn	A	11 32 13	Yugoslavia 45.45 N 14.28 E
	ePg	A	32 36	H = 11 30 51.7 h = 22 km
	iSn	A	33 12	D = 5.50 Az = 342.0 (NEIS)
	e	A	33 40	
	eSg	A	33 44	
5.	ePKP	A	19 49 46	Tonga Islands Region 17.35 S 172.77 W
	e	A	49 53	H = 19 30 07.0 h = normal MB = 5.0
				D = 146.60 Az = 354.9 (NEIS)
				PKPV A 1.4s 32.6nm
5.	eP	A	21 32 22.5	Kurile Islands 43.41 N 146.72 E
	e	A	32 32.5	H = 21 20 23.4 h = normal MB = 5.0
				D = 78.50 Az = 332.7 (NEIS)
				PV A 1.1s 16.1nm M = 4.9

Day	Phase		h m s	Remarks
5.	LmH	B	21 42.0	LmH B 18s 0.4/um
	LmV	B	42.0	LmV B 18 0.4/um
6.	LmV	B	05 33.7	Off Coast of Central Chile
	LmH	B	34.5	33.07 S 72.24 W
				H = 04 24 50 h = 25 km MB = 4.9 (ISC)
				D = 111.1
				LmH B 17s 0.6/um M = 5.3
				LmV B 17 0.7/um 5.3
6.	LmH	C	07 48.5	Near Coast of Central Chile
	LmV	C	48.5	31.73 S 71.95 W
				H = 06 43 37.0 h = 42 km MB = 5.0 (ISC)
				D = 110.0
				LmH C 17s 0.6/um M = 5.2
				LmV C 18 0.55/um 5.1
6.	eP dif	B	15 22 32	Southwestern Atlantic Ocean
	ePKIKP	A	26 17	60.82 S 21.55 W
	ei	A	26 31	H = 15 07 37.3 h = normal MB=6.2 MS=7.0
	ePP	AB	27 14	D = 114.27 Az = 22.5 (NEIS)
	ePS	B	37 00	PPV A 1.6s 55.0nm M = 5.9
	ePKKP	A	37 04	PPV B 5.5 1.9/um 6.9
	ePcPPKP	A	40 57	LmH B 20 25.1/um 6.6
	e	A	41 12.5	LmV B 20 20.6/um 6.7
	e	A	41 19	
	eiSS	B	43 16	
	LmV	B	16 10.4	
	LmH	B	10.5	
6.	e(P)	A	16 48 03	Mindanao, Philippine Islands
				9.73 N 125.62 E
				H = 16 34 33.3 h = 82 km MB = 5.0
				D = 97.23 Az = 324.1 (NEIS)
6.	eP	A	21 24 22	Crete 34.87 N 26.25 E
	LmH	B	32.9	H = 21 19 58.2 h = 33 km MB = 4.4

Day	Phase		h m s	Remarks
cont. 6.	LmV	B	21 33.0	D = 19.02 Az = 330.4 (NEIS) PV A 2.0s 25.6nm M = 4.1 LmH B 12 0.25/um 3.8 LmV B 12 0.5/um 4.2
7.	eSg	A	02 50 37.5	Poland 50.25 N 18.92 E H = 02 48 05.5 M = 2.6 (WAR) D = 4.7
7.	e	A	07 17 47	
7.	ePKP	A	07 56 22	Samoa Islands Region 15.97 S 172.78 W H = 07 36 45.1 h = normal MB = 4.9 D = 145.23 Az = 355.1 (NEIS) PKPV A 1.5s 50.2nm
7.	eP	A	09 39 07.5	Off Coast of Hokkaido, Japan
	e	A	39 16	42.34 N 146.61 E
	e	A	39 34.5	H = 09 27 02.2 h = 27 km MB=5.2 MS=4.7
	eS	B	49 12	D = 79.41 Az = 332.7 (NEIS)
	LmH	B	10 20.3	PV A 1.4s 23.2nm M = 4.9
	LmV	B	21.0	LmH B 12.5 0.6/um 5.2 LmV B 12 0.6/um 5.2
7.	LmH	B	12 07.6	Luzon 16.32 N 122.32 E
	LmV	B	11.0	H = 11 11 24.3 h = 57 km MB = 4.9 (ISC) D = 90.1 LmH B 16s 0.25/um LmV B 16 0.35/um
7.	LmH	B	13 17.0	Luzon 16.22 N 122.36 E
	LmV	B	20.0	H = 12 20 44.8 h = 106 km MB = 4.8 (ISC) D = 90.3 LmH B 15.5s 0.3/um LmV B 16 0.4/um
8.	e	A	01 17 52	

Day	Phase		h m s	Remarks
8.	ePKIKP	A	04 04 48	Fiji Islands Region 17.79 S 178.26 W
	ePKHKP	A	04 50	H = 03 46 09.0 h = 544 km MB = 5.4
				D = 146.28 Az = 348.7 (NEIS)
8.	eP	A	05 38 22	Kurile Islands 45.66 N 151.65 E
				H = 05 26 27.6 h = 60 km MB = 4.5
				D = 78.03 Az = 335.3 (NEIS)
8.	ePKHKP	A	06 37 52	South of Fiji Islands 25.03 S 179.64 E
	e	A	38 12	H = 06 18 51.8 h = 495 km MB = 4.6
				D = 152.81 Az = 343.2 (NEIS)
				traces
8.	eP	A	08 38 15	Kurile Islands 45.79 N 151.73 E
				H = 08 26 20.6 h = 60 km MB = 4.5
				D = 77.93 Az = 335.3 (NEIS)
9.	eP	A	01 57(50)	Luzon, Philippine Islands
	eS	B	02 08 45	16.29 N 122.36 E
	LmH	B	41.3	H = 01 44 57.5 h = normal
	LmV	B	44.3	MB = 5.2 MS = 5.1 (NEIS)
				D = 90.09
				PV A 1.8s 13.5nm M = 5.0
				LmH B 15.3 1.9/um 5.6
				LmV B 15 2.4/um 5.8
9.	eP	A	07 59 15	Kurile Islands 43.30 N 146.91 E
				H = 07 47 14.9 h = normal MB = 5.1 MS=5.3
				D = 78.66 Az = 332.8 (NEIS)
				PV A 1.1s 14.1nm M = 4.9
9.	ePKP	A	08 17 04	New Hebrides Islands 14.23 S 167.23 E
	ePP	B	19 56	H = 07 57 31.0 h = 9 km MB=5.8 MS=6.4
	LmH	B	09 15.7	D = 138.65 Az = 336.6 (NEIS)
	LmV	B	21.5	PPV B 9s 1.1/um M = 6.1
				LmH B 18.3 6.5/um 6.4
				LmV B 19 4.8/um 6.3

Day	Phase		h m s	Remarks
9.	eP	A	19 15 52.5	Central Mid-Atlantic Ridge
	LmV	B	36.9	7.43 N 35.17 W
	LmH	B	39.2	H = 19 06 02.0 h = normal MB = 4.9
				D = 57.90 Az = 33.2 (NEIS)
				PV A 1.2s 20.3nm M = 5.1
9.	eSg	A	23 03 45	Merlebach, France 49.17 N 6.78 E
				H = 23 01 52.0 h = 0 km (BCIS)
				D = 3.92 Az = 62 (ISC)
9.	LmH	C	23 43.0	Near Coast of Central Chile
	LmV	C	43.9	31.71 S 71.95 W
				H = 22 38 49.7 h = 39 km MB=5.2 (ISC)
				D = 110.0
				LmH C 19s 0.35/um M = 4.9
				LmV C 19 0.4/um 5.0
10.	eP	A	11 10 11	Eastern Mediterranean Sea
				34.40 N 28.31 E
				H = 11 05 35.5 h = 63 km MB = 4.5
				D = 20.20 Az = 328.2 (NEIS)
10.	LmH	B	20 11.0	Gulf of California 29.5 N 113.6 W
	LmV	B	11.1	H = 19 16 54 h = 36 km (ISC)
				D = 86.7
				LmH B 17s 0.25/um M = 4.7
				LmV B 17 0.3/um 4.8
11.	eP	A	00 14 05	Gulf of California 29.72 N 113.47 W
				H = 00 01 24.1 h = normal MB = 5.0
				D = 86.45 Az = 31.5 (NEIS)
				PV A 2.0s 25.6nm M = 5.1
11.	eP	AB	02 17 49	Central Mid-Atlantic Ridge
	e	A	18 02	0.83 N 29.21 W
	e(PPP)	B	21 48	H = 02 07 40.5 h = normal MB = 5.1
	ePS	BC	26 28	D = 60.48 Az = 28.6 (NEIS)
	LmH	B	42.3	PV A 2.0s 59.8nm M = 5.4

Day	Phase		h m s	Remarks
cont. 11.	LmV	B	02 42.4	SH B 19s 11.3 µm M = 6.4
	P'P'	A	47 37	LmH B 18 8.9 µm 5.9
	e	A	47 43	LmV B 17 8.7 µm 6.0
11.	ePKP	A	08 57 32	<u>Fiji Islands Region</u> 17.99 S 178.74 W
	e	A	57 38.5	H = 08 39 00.8 h = 654 km MB = 4.7
				D = 146.39 Az = 348.1 (NEIS)
11.	LmH	B	10 56.8	<u>West Irian Region</u> 4.61 S 134.03 E
	LmV	B	03.3	H = 09 53 35.0 h = 33 km (ISC)
				D = 113.7
				LmH B 18s 1.7 µm M = 5.7
				LmV B 18 0.5 µm 5.2
11.	LmH	B	14 09.3	<u>Gulf of California</u> 29.6 N 113.7 W
	LmV	B	11.0	H = 13 16 42 h = 0 km (ISC)
				D = 86.7
				LmH B 17s 0.25 µm M = 4.7
				LmV B 15 0.45 µm 5.0
11.	eP	A	18 45 03	<u>Off East Coast of Honshu, Japan</u>
	epP	A	45 14	34.88 N 141.56 E
	LmH	B	19 24.8	H = 18 32 35.4 h = 52 km MB = 4.6
	LmV	B	28.2	D = 84.10 Az = 330.6 (NEIS)
				h = 40 km
				pPV A 1.4s 11.6nm
				LmH B 16 1.6 µm
				LmV B 16 0.7 µm
11.	eP	A	19 46 44	<u>Gulf of California</u> 29.71 N 113.38 W
				H = 19 34 04.7 h = 32 km MB = 4.9
				D = 86.41 Az = 31.5 (NEIS)
				PV A 1.4s 9.3nm M = 4.8
11.	ePKHKP	A	21 38 18.5	<u>Fiji Islands Region</u> 18.31 S 178.08 W
	ePKP2	A	38 21.5	H = 21 19 42.8 h = 619 km MB = 5.1
				D = 146.83 Az = 348.7 (NEIS)
				PKHKPV A 1.3s 19.7nm

Day	Phase		h m s	Remarks
12.	eP1	A	03 02 16.5	<u>Tadzhik SSR</u> 37.75 N 71.95 E
	eP2	AB	02 19	H = 02 54 07.7 h = 11 km MB=5.3 MS=4.6
	ePP	A	04 06	D = 43.96 Az = 307.2 (NEIS)
	LmH	B	18.9	P2V A 1.7s 54.6nm M = 5.1
	LmV	B	22.9	LmH B 28 2.4 μm 5.0
				LmV B 11 1.0 μm 5.1
12.	eP	A	06 06 38	<u>Off Coast of Oregon</u> 43.74 N 127.47 W
	LmV	B	46.2	H = 05 54 27.7 h = 6 km MB=5.4 MS=5.2
	LmH	B	46.7	D = 79.49 Az = 25.1 (NEIS)
				PV A 1.6s 27.5nm M = 5.0
				LmH B 14.7 1.0 μm 5.2
				LmV B 16 0.9 μm 5.3
12.	eP	A	17 38 03	<u>Southern Nevada</u> 37.20 N 116.20 W
				H = 17 00 00.8 h = 0 km MB = 4.8
				D = 81.23 Az = 30.6 (NEIS)
				Nuclear explosion "HUSKY ACE" (AEC)
12.	LmH	B	19 02.0	<u>Near Coast of Peru</u> 16.02 S 74.05 W
	LmV	B	02.0	H = 18 04 29.3 h = 47 km MB=5.3 MS=4.6
				D = 99.51 Az = 40.1 (NEIS)
				LmH B 19s 0.5 μm M = 5.0
				LmV B 18 0.7 μm 5.2
12.	eP	A	23 12 19	<u>North Atlantic Ridge</u> 12.02 N 43.66 W
				H = 23 02 20.6 h = normal MB = 4.9
				D = 59.05 Az = 37.6 (NEIS)
				PV A 1.2s 16.3nm M = 5.0
12.	ePKP	A	23 56 37	<u>Fiji Islands Region</u> 17.55 S 179.15 W
				H = 23 37 59.9 h = 588 km MB = 4.8
				D = 145.88 Az = 347.8 (NEIS)
13.	eP	A	01 56 30.5	<u>Gulf of California</u> 29.58 N 113.64 W
	LmV	B	02 36.5	H = 01 43 47.2 h = 15 km MB = 5.2
	LmH	B	38.0	D = 86.65 Az = 31.4 (NEIS)
				LmH B 16s 0.3 μm M = 4.8
				LmV B 16 0.6 μm 5.1

Day	Phase		h m s	Remarks
13.	eP	A	06 04 56	Crete 34.73 N 26.18 E
	e	A	05 02	H = 06 00 33.7 h = 39 km MB = 4.6
				D = 19.12 Az = 330.7 (NEIS)
13.	eP	A	16 36 20	Hokkaido, Japan Region 41.88 N 142.29 E
				H = 16 24 27.3 h = 71 km MB = 4.9
				D = 78.29 Az = 330.4 (NEIS)
13.	eP	A	20 37 06	Kyushu, Japan 30.69 N 131.94 E
	eX	A	37 16.5	H = 20 24 37.2 h = 12 km MB = 5.1
				D = 83.41 Az = 326.4 (NEIS)
				PV A 2.0s 29.9nm M = 5.1
				XV A 1.6 38.5nm
13.	eP	A	21 20 52	Central Mid-Atlantic Ridge
	LmH	B	44.7	4.80 N 32.59 W
	LmV	B	46.2	H = 21 10 49.3 h = normal MB = 5.4
				D = 58.74 Az = 31.3 (NEIS)
				PV A 2.0s 72.6nm M = 5.4
				LmH B 18 0.6 µm 4.7
				LmV B 17 0.9 µm 5.0
13.	eP	A	21 22 35	Gulf of California 29.43 N 113.61 W
	LmH	B	22 00.5	H = 21 09 51.7 h = normal MB = 4.7
	LmV	B	00.6	D = 86.75 Az = 31.4 (NEIS)
				LmH B 18s 0.7 µm M = 5.1
				LmV B 20 0.9 µm 5.2
13.	LmH	B	23 04.7	Central Mid-Atlantic Ridge
	LmV	B	06.3	5.09 N 32.65 W
				H = 22 30 48.3 h = 33 km MB = 4.9 (ISC)
				D = 58.5
				LmH B 18s 0.5 µm M = 4.7
				LmV B 16 0.7 µm 4.9
14.	LmV	B	01 34.0	Mid-Atlantic Ridge 4.79 N 32.58 W
				H = 00 55 39.5 h = 33 km MB = 4.8 (ISC)
				D = 58.8
				LmV B 18s 0.45 µm M = 4.7

Day	Phase		h m s	Remarks
14.	eP	A	06 32 24.5	North Atlantic Ocean 58.60 N 31.87 W
				H = 06 26 54.4 h = normal MB = 4.5
				D = 25.99 Az = 88.7 (NEIS)
				PV A 1.5s 20.1nm M = 4.5
14.	eP	A	10 06 10	Central Mid-Atlantic Ridge
	LmH	B	29.8	5.09 N 32.69 W
	LmV	B	29.9	H = 09 56 14.8 h = normal MB = 5.0
				D = 58.55 Az = 31.4 (NEIS)
				LmH B 19s 0.3/um M = 4.4
				LmV B 19 0.35/um 4.6
14.	eP1	A	18 11 28	Crete 34.75 N 26.23 E
	eP2	AC	11 33	H = 18 07 05.7 h = 45 km MB = 4.8
	eS	B	15 04	D = 19.12 Az = 330.6 (NEIS)
	LmH	B	20.9	P2V A 1.7s 48.5nm M = 4.4
	LmV	B	20.9	LmH B 13.5 2.7/um 4.7
				LmV B 14 2.4/um 4.8
14.	eP	A	19 43 58	North of Severnaya Zemlya
				84.8 N 101 E
				H = 19 36 21 h = 0 km MB = 4.6
				D = 39.81 Az = 276 (ISC)
14.	eP	AB	22 15 16	North of Severnaya Zemlya
	e	A	15 24.5	85.05 N 99.48 E
	eS	B	21 20	H = 22 07 46.8 h = normal MB=5.2 MS=5.3
	LmV	B	38.0	D = 39.62 Az = 273.9 (NEIS)
	LmH	B	38.2	PV A 1.4s 69.7nm M = 5.2
				LmH B 15 1.2/um 4.8
				LmV B 14 1.3/um 5.0
14.	ePKIKP	A	22 25 49	Solomon Islands 6.36 S 154.90 E
	e	A	26 02.5	H = 22 06 50.4 h = 60 km MB = 5.4
				D = 126.30 Az = 331.8 (NEIS)
15.	ePKP2	A	02 23 38	Kermadec Islands 30.0 S 178.1 W
				H = 02 03 11.6 h = 22 km
				D = 158.07 Az = 343 (ISC)

Day	Phase		h m s	Remarks
15.	eP	A	03 33 47.5	Kurile Islands 45.38 N 151.42 E
				H = 03 21 50 h = 19 km MB = 4.5
				D = 78.21 Az = 335 (ISC)
15.	ePKHKP	BC	06 44 44	South of Kermadec Islands
	ePKP2	A	45 06	32.77 S 179.80 W
	ePP	BC	48 56	H = 06 24 39 h = 59 km MB = 5.1
	LmH	B	07 56.0	D = 160.25 Az = 338 (ISC)
	LmV	B	57.0	LmH B 20s 0.4 μm
				LmV B 20 0.4 μm
16.	ePKP2	A	00 39 11.5	South of Kermadec Islands
				32.62 S 179.68 W
				H = 00 18 35 h = 21 km
				D = 160.15 Az = 338 (ISC)
				PKP2V A 2.0s 34.2nm
16.	eP	A	10 00 29.5	Nepal 28.22 N 82.95 E
				H = 09 50 43.3 h = normal MB = 5.2
				D = 57.14 Az = 314.1 (NEIS)
				PV A 1.1s 16.1nm M = 5.0
16.	eP	A	17 04 02	Fox Islands, Aleutian Is.
				52.36 N 168.33 W
				H = 16 52 08.6 h = 32 km MB = 4.3
				D = 77.37 Az = 0.0 (NEIS)
17.	+eP	AB	03 24 07.5	Afghanistan - USSR Border Region
	esP	AB	25 16	36.37 N 71.16 E
	ePP	B	25 59	H = 03 16 18.6 h = 221 km MB = 5.5
	e	B	26 16	D = 44.30 Az = 308.2 (NEIS)
	epPP	B	26 32	h = 215 km
	esPP	B	27 00	PV A 1.9s 326.0nm M = 5.4
	eS	B	30 30	PV B 8 1.5 μm 5.5
	esS	B	31 44	
	eScS	B	33 44	
	eiSS	B	33 56	

Day	Phase		h m s	Remarks
17.	eP	A	20 37 21	<u>Panama - Colombia Border Region</u>
	LmV	C	21 07.6	7.50 N 77.25 W
	LmH	C	08.0	H = 20 24 52.2 h = 15 km MB = 4.9
				D = 83.54 Az = 39.8 (NEIS)
				PV A 1.8s 40.5nm M = 5.4
				LmH C 26 0.3/um 4.6
				LmV C 27 0.4/um 4.7
17.	eP	A	21 14 45.5	<u>Chiapas, Mexico</u> 17.36 N 93.80 W
	esP	A	18 28	H = 21 02 24.9 h = 202 km MB = 4.7
				D = 86.16 Az = 38.0 (NEIS)
				h = 214 km
				PV A 1.2s 20.3nm M = 4.8
17.	e	A	22 59 08	<u>South Atlantic Ridge</u> 25.44 S 13.72 W
				H = 22 46 59.9 h = normal MB = 4.5
				D = 79.00 Az = 16.1 (NEIS)
18.	eP	AB	01 21 45	<u>Costa Rica</u> 9.17 N 83.98 W
	eS	B	32 18	H = 01 09 04.8 h = 41 km MB=5.3 MS=5.1
	eSS	B	38 13	D = 86.52 Az = 39.4 (NEIS)
	eSSS	C	41 48	PV A 1.5s 65.4nm M = 5.6
	LmV	B	02 00.0	LmH B 18.5 0.9/um 5.2
	LmH	B	00.1	LmV B 18 1.1/um 5.4
18.	+eP	AB	11 02 37.5	<u>Near Coast of Jalisco, Mexico</u>
	e	A	02 44	19.38 N 104.98 W
	ePP	B	06 12	H = 10 49 37.5 h = 45 km MB=6.0 MS=5.5
	eSKS	B	13 12	D = 90.84 Az = 34.7 (NEIS)
	ePS	B	14 44	PV A 2.4s 262.0nm M = 6.2
	eSS	C	20 20	PV B 8 1.1/um 6.2
	LmV	B	48.1	LmH B 13.7 3.6/um 6.0
	LmH	B	49.9	LmV B 14 6.0/um 6.2
18.	-eiP	●A	11 47 53	<u>North of Severnaya Zemlya</u>
				84.97 N 99.64 E
				H = 11 40 23.4 h = normal MB = 4.9
				D = 39.64 Az = 274.2 (NEIS)
				PV A 1.9s 114.0nm M = 5.3

Day	Phase		h m s	Remarks
18.	LmV	B	18 49.8	Near Coast of Northern Chile
	LmH	B	53.0	26.78 S 70.8 W
				H = 17 46 15.6 h = 42 km MB = 4.7 (ISC)
				D = 105.7
				LmH B 16s 0.25 /um M = 4.8
				LmV B 19 0.4 /um 5.0
19.	ePP	A	00 20 02	Southern Greece 36.93 N 21.57 E
				H = 00 16 04.5 h = normal MB = 4.4
				D = 15.46 Az = 335.6 (NEIS)
19.	ePKIKP	A	00 32(55)	Macquarie Islands Region 54.69 S 158.54 E
	ePKP2	AB	33 34.5	H = 00 13 00.7 h = 33 km MB = 5.5
	ePP	AB	37 14	D = 159.65 Az = 272 (ISC)
	LmH	C	01 56.2	PKP2V A 1.5s 50.2nm
	LmV	C	02 07.4	LmH C 20 0.3 /um M = 4.9
				LmV C 16 0.5 /um 5.3
19.	eP	A	15 51 14	Greenland Sea 74.05 N 9.0 E
				H = 15 46 05.4 h = 33 km
				D = 23.52 Az = 176 (ISC)
				PV A 1.4s 14.0nm M = 4.3
19.	eP	A	21 16 06	Kurile Islands 44.09 N 148.95 E
	epP	A	16 20	H = 21 04 07.8 h = 51 km MB = 4.7
	e	A	16 23	D = 78.62 Az = 333.9 (NEIS)
				h = 53 km
19.	eP	A	23 52 06.5	Kurile Islands 44.18 N 149.04 E
	e	A	52 21.5	H = 23 40 09.4 h = 51 km MB = 5.2
	epP	A	52 24.5	D = 78.56 Az = 333.9 (NEIS)
				h = 67 km
				PV A 1.4s 30.2nm M = 5.1
20.	eP	A	00 36 34.5	Kurile Islands 44.07 N 149.11 E
				H = 00 24 36.3 h = 49 km MB = 4.7
				D = 78.69 Az = 334.0 (NEIS)
				PV A 1.5s 20.1nm M = 4.9

Day	Phase		h m s	Remarks
20.	e	A	01 15 54	
20.	LmH	C	11 09.5	Central Mid-Atlantic Ridge (ISC)
	LmV	C	10.1	LmH C 24s 0.2/um
				LmV C 22 0.3/um
20.	ePKP	A	16 29 55	Fiji Islands Region 18.33 S 178.24 W
				H = 16 11 17.5 h = 596 km MB = 4.5
				D = 146.82 Az = 348.5 (NEIS)
				PKPV A 1.6s 16.5nm
20.	eP	A	18 19 48.5	Hokkaido, Japan Region 41.65 N 143.73 E
				H = 18 07 47.1 h = 38 km MB = 4.7
				D = 79.01 Az = 331.2 (NEIS)
				traces
20.	ePKIKP	A	20 11 43	Fiji Islands Region 21.69 S 179.38 W
	eiPKHKP	A	11 48.5	H = 19 53 05.5 h = 614 km MB = 5.4
	ePKP2	A	11 55.5	D = 149.83 Az = 346.0 (NEIS)
	epPKP	A	14 11	PKIKPV A 1.8s 27.0nm
				PKHKPV A 1.4 112.0nm
				PKP2V A 1.7 48.5nm
21.	eP	A	11 11 37.5	Unimak Island Region 53.80 N 163.C7 W
	epP	A	11 48	H = 10 59 53.1 h = 41 km MB = 4.8
				D = 75.84 Az = 3.5 (NEIS)
				h = 39 km
21.	ePKHKP	A	11 40 34.5	South of Fiji Islands 22.15 S 179.64 W
	ePKP2	A	40 44.5	H = 11 21 50.8 h = 607 km MB = 5.0
	epPKP	A	42 55	D = 150.22 Az = 345.5 (NEIS)
21.	ePKIKP	A	13 58 02	Fiji Islands Region 21.17 S 178.69 W
	ePKHKP	A	58 07	H = 13 39 20.7 h = 572 km MB = 5.0
	ePKP2	A	58 14	D = 149.48 Az = 347.0 (NEIS)
				PKIKPV A traces

Day	Phase		h m s	Remarks
21.	ePKP2	A	14 22 40	South of Fiji Islands 24.14 S 176.45 W
	ePP	A	26 26	H = 14 02 45.2 h = 85 km MB = 5.1
				D = 152.81 Az = 348.7 (NEIS)
				PKP2V A 1.0s 11.8nm
21.	eP	A	19 52 39.5	Near East Coast of Honshu, Japan
				35.58 N 140.71 E
				H = 19 40 18.5 h = 60 km MB = 4.7
				D = 83.14 Az = 330.2 (NEIS)
21.	ePKIKP	A	21 27 02.5	South of Fiji Islands 23.84 S 179.66 E
	ePKHKP	A	27 09.5	H = 21 08 17.8 h = 558 km MB = 4.3
				D = 151.68 Az = 343.9 (NEIS)
23.	eP	A	10 53 33.4	Rumania 45.69 N 26.50 E
	e	A	54 03	H = 10 50 58.6 h = 174 km MB = 4.9
	LmV	B	11 11.0	D = 11.10 Az = 301.9 (NEIS)
	LmH	B	13.0	PV A 1.3s 13.1nm M = 4.3
				LmH B 15.5 0.8/um
				LmV B 15 1.0/um
23.	LmV	B	22 24.6	LmH B 17s 1.1/um
	LmH	B	25.1	LmV B 16 0.9/um
24.	e	A	01 44 12	
24.	ePKP	A	03 27 57	Fiji Islands Region 16.80 S 177.11 W
				H = 03 08 19.2 h = normal MB = 5.0
				D = 145.51 Az = 350.2 (NEIS)
				PKPV A 1.5s 15.1nm
24.	eP	A	04 38 43	Sea of Okhotsk 48.34 N 146.20 E
				H = 04 27 54.8 h = 470 km MB = 4.6
				D = 73.97 Az = 331.8 (NEIS)
				traces

Day	Phase		h m s	Remarks
24.	eP	A	04 54 42.5	Near East Coast of Kamchatka
				56.04 N 162.91 E
				H = 04 43 26.6 h = normal MB = 4.7
				D = 71.04 Az = 341.1 (NEIS)
				PV A 1.4s 14.0nm M = 4.8
24.	eP	A	04 59 57.5	Kirgiz-Sinkiang Border Region
				41.54 N 79.46 E
				H = 04 51 33.0 h = normal MB = 4.6
				D = 46.39 Az = 305.5 (NEIS)
				traces
24.	eP	A	05 32 39.5	Eastern Kashmir 33.11 N 75.89 E
	epP	A	32 47.5	H = 05 23 51.0 h = normal MB = 5.4
	LmH	B	56.3	D = 49.37 Az = 310.9 (NEIS)
				h = 34 km
				PV A 1.5s 55.3nm M = 5.4
24.	eP	A	20 06 04	Eastern Kashmir 33.06 N 75.73 E
	epP	A	06 14	H = 19 57 17.9 h = 52 km MB = 5.0
				D = 49.30 Az = 310.9 (NEIS)
				h = 44 km
24.	e(PKP)	A	23 19 43	Loyalty Islands 19.98 S 168.56 E
				H = 23 00 15.3 h = 140 km MB = 4.4
				D = 144.38 Az = 334.7 (NEIS)
25.	eP	A	01 21 54.5	Rumania 45.55 N 26.52 E
				H = 01 19 16.6 h = 137 km MB = 3.8
				D = 11.19 Az = 302.5 (NEIS)
				PV A 1.3s 10.9nm M = 4.4
25.	eP	AB	06 54 09	Mindoro, Philippine Islands
	eSKS	B	07 04 37	13.80 N 120.24 E
	eS	B	04 56	H = 06 41 11.3 h = 63 km MB = 5.6
	ePS	B	06 08	D = 90.85 Az = 322.9 (NEIS)
	LmH	B	39.6	PV A 1.8s 43.9nm M = 5.5
	LmV	B	40.4	LmH B 16.3 3.5 /um
				LmV B 16 3.6 /um

Day	Phase		h m s	Remarks
25.	-iP	AB	14 21 39	Salta Province, Argentina
	+ipP	AB	23 37	21.99 S 63.66 W
	esPPP	B	30 23	H = 14 08 59.5 h = 529 km MB = 6.1
	iSKS	AB	31 29	D = 97.86 Az = 38.4 (NEIS)
	eS	B	32 19	PV A 1.8s 189.0nm M = 6.1
	eSP	B	33 41	PV B 8 1.36 um 6.3
	ePS	B	34 58	PKKPV A 1.4 46.5nm
	esS	B	35 55	e 14 36 51 e 38 10.5 e 40 14.5
	eisSP	B	37 16	
	ePKKP	A	38 42.5	
	eSS	B	38 56	
	eSSS	B	42 15	
25.	e(sP)	A	15 03 49.5	Peru 13.62 S 72.44 W
				H = 14 49 51.3 h = 107 km MB = 5.1 (ISC)
				D = 96.7
25.	epP	A	15 25 19.5	Southern Bolivia 21.76 S 63.56 W
				H = 15 10 39.2 h = 532 km MB = 5.2 (ISC)
				D = 97.7
26.	eP	A	03 34 24.5	Near East Coast of Kamchatka
				56.27 N 162.86 E
				H = 03 23 09.2 h = normal MB = 4.4
				D = 70.81 Az = 341.1 (NEIS)
26.	+iP	A	04 34 47	Eastern Kazakh SSR 49.77 N 78.20 E
	ePn	A	36 20	H = 04 26 57.7 h = 0 km MB=5.3 MS=4.4
				D = 41.31 Az = 297.8 (NEIS)
				Underground explosion MB = 5.7 (UPP)
				PV A 0.8s 53.8nm M = 5.3
26.	eP	A	05 51 49	Southeastern Alaska 59.11 N 135.37 W
				H = 05 40 56.1 h = normal MB = 4.8
				D = 67.35 Az = 22.1 (NEIS)
				PV A 1.2s 18.3nm M = 5.1

Day	Phase		h m s	Remarks
26.	eP	A	06 05 40.5	Ural Mountains Region 53.66 N 55.38 E
				H = 05 59 57.6 h = 0 km MB = 4.8
				D = 26.70 Az = 281.5 (NEIS)
				Underground explosion MB = 4.9 (UPP)
26.	eP	A	11 23 40	Off Coast of Jalisco, Mexico
	LmH	B	12 05.0	17.84 N 106.08 W
	LmV	B	06.6	H = 11 10 28.7 h = normal MB=5.3 MS=5.1
				D = 92.70 Az = 34.4 (NEIS)
				LmH B 16s 0.8/um M = 5.3
				LmV B 16 1.1/um 5.4
26.	ePKHKP	A	15 30 49	Fiji Islands Region 20.37 S 177.73 W
				H = 15 11 59.1 h = 525 km MB = 4.7
				D = 148.89 Az = 348.5 (NEIS)
26.	pP	A	18 48 40	South of Honshu, Japan 29.00 N 142.41 E
	LmV	B	19 34.7	H = 18 35 33.4 h = 43 km MB=5.1 MS=4.6
				D = 89.57 Az = 331.2 (NEIS)
				h = 44 km
				LmV B 14s 0.35/um M = 5.0
26.	eP	A	20 25 15	Greece 38.72 N 20.95 E
	e	A	25 25	H = 20 22 03.6 h = 55 km MB = 3.9
				D = 13.64 Az = 334.0 (NEIS)
				PV A traces
27.	eP	A	00 23 38	Mediterranean Sea 35.1 N 14.7 E
				H = 00 19 50 (BCIS)
				D = 15.80
27.	+iP	AB	07 05 51	Novaya Zemlya 70.78 N 54.18 E
	eiPn	B	06 18	H = 06 59 57.4, h = 0 km MB=6.9 MS=5.5
	eiPP	B	06 34	D = 28.00 Az = 246.5 (NEIS)
	eiS	B	10 38	Underground explosion MB = 6.9 (UPP)
	iL	BH	14 28	PV A 1.2s 1990.0nm M = 6.8
	iL	BV	14 46	PV B 3.0 7.36/um 7.0

Day	Phase		h m s	Remarks
cont. 27.	LmV	B	07 17.7	LmH B 12s 50.0/um M = 6.3
	LmH	B	17.8	LmV B 12 54.7/um 6.5
27.	ePKP	A	08 13 16	New Hebrides Islands 17.98 S 169.09 E H = 07 53 38.2 h = 29 km MB = 5.3 D = 142.78 Az = 336.2 (NEIS)
27.	eP	A	09 19 42	Novaya Zemlya 71.30 N 51.88 E H = 09 13 51.3 h = 0 km MB = 4.8 D = 27.55 Az = 242.8 (NEIS)
27.	e	A	09 25 03	
27.	eP	A	09 59 05	Near Coast of West Pakistan 24.63 N 62.14 E H = 09 50 38.2 h = normal MB = 4.7 D = 46.53 Az = 317.4 (NEIS) PV A 1.0s 9.8nm M = 4.7
27.	eP	A	15 50 38	Arabian Sea 14.35 N 53.32 E H = 15 41 48.8 h = normal MB = 4.7 D = 49.41 Az = 326.1 (NEIS)
28.	ePn	A	03 57 22	Czechoslovakia 48.42 N 17.08 E
	ePg	A	57 37.5	H = 03 56 15.4 h = 2 km
	eSn	A	58 12	D = 4.20 Az = 304.1 (NEIS)
	eSg	A	58 28.5	
28.	eP	A	10 06 51.5	Iceland Region 66.60 N 19.88 W H = 10 01 54.1 h = normal MB = 4.5 D = 22.51 Az = 119.7 (NEIS) PV A 1.8s 33.8nm M = 4.5
28.	e	A	10 17 04	
28.	eP	A	10 47 49	Iceland Region 66.75 N 19.66 W H = 10 42 50.6 h = normal MB = 4.4 D = 22.51 Az = 120.3 (NEIS)

Day	Phase		h m s	Remarks
28.	eP	A	10 58 18	Iceland Region 66.83 N 19.68 W
				H = 10 53 21.3 h = normal MB = 4.3
				D = 22.56 Az = 120.5 (NEIS)
				PV A 1.7s 15.2nm M = 4.2
28.	eP	A	11 16 58.5	Iceland Region 66.99 N 19.33 W
	LmV	B	26.5	H = 11 12 02.5 h = normal MB=4.7 MS=3.9
	LmH	B	29.2	D = 22.52 Az = 121.3 (NEIS)
				PV A 1.8s 47.4nm M = 4.7
				LmH B 12 0.6 µm 4.3
				LmV B 16 0.5 µm 4.2
28.	eP	A	11 20 27.5	Iceland Region 67.07 N 19.4 W
				H = 11 15 25.0 h = 0 km MB = 4.3
				D = 22.59 Az = 121 (ISC)
28.	eP	A	11 30 38	Iceland Region 66.82 N 19.54 W
				H = 11 25 40.6 h = 33 km MB = 4.4 (NEIS)
				D = 22.51
28.	eP1	AB	11 36 41.5	Iceland Region 66.95 N 19.22 W
	eP2	B	36 47	H = 11 31 44.1 h = normal MB=5.0 MS=4.4
	eS	B	40 52	D = 22.47 Az = 121.4 (NEIS)
	LmH	B	45.4	P1V A 1.8s 74.4nm M = 4.8
	LmV	B	47.6	P2V B 4.5 0.55 µm 5.3
				SH B 11 0.91 µm 5.0
				LmH B 18 1.2 µm 4.4
				LmV B 13 1.4 µm 4.7
28.	eP1	A	11 52 33	Iceland Region 66.84 N 19.33 W
	eP2	A	52 36	H = 11 47 37.6 h = normal MB = 4.7
				D = 22.45 Az = 121.0 (NEIS)
				P2V A 1.7s 36.4nm M = 4.7
28.	eP	A	12 06 46.5	Iceland Region 67.30 N 18.99 W
				H = 12 01 47.8 h = normal MB = 4.4
				D = 22.58 Az = 122.4 (NEIS)

Day	Phase		h m s	Remarks
28.	ePKIKP	A	14 04 37	South of Kermadec Islands
	ePKP2	A	05 25	33.04 S 179.15 W
				H = 13 44 45.9 h = 95 km MB = 5.5
				D = 160.70 Az = 338.9 (NEIS)
28.	eP	A	14 30 51	Iceland Region 66.91 N 19.47 W
	LmH	B	39.6	H = 14 25 54.3 h = normal MB=4.5 MS=4.0
	LmV	B	40.3	D = 22.53 Az = 120.9 (NEIS)
				PV A 1.2s 16.3nm M = 4.4
				LmH B 19 0.6 μm 4.0
				LmV B 18 0.7 μm 4.3
28.	eP	A	15 04 36.5	Negros, Philippine Islands
				9.12 N 123.93 E
				H = 14 52 04.1 h = 554 km MB = 5.1
				D = 96.74 Az = 323.6 (NEIS)
				PV A 1.0s 13.8nm M = 5.2
28.	eP	A	22 34 34.5	Near East Coast of Honshu, Japan
				35.60 N 140.75 E
				H = 22 22 12.9 h = 57 km MB = 4.9
				D = 83.14 Az = 330.2 (NEIS)
29.	ePP	A	05 46 58	West New Guinea 2.98 S 139.27 E
	e	A	47 09	H = 05 27 25.50 h = 36 km
	e	A	47 21	MB = 5.7 MS = 5.5 (NEIS)
	e	A	47 30	D = 115.35
	LmH	B	06 39.4	PPV A 2.5s 61.5nm M = 5.9
	LmV	B	39.7	LmH B 20 1.9 μm 5.7
				LmV B 19 2.4 μm 5.8
29.	eP	A	08 46 42	Iceland Region 66.50 N 19.29 W
				H = 08 41 47.0 h = normal MB = 4.3
				D = 22.26 Az = 120.3 (NEIS)
29.	eP	A	14 07 28	Iceland Region 66.50 N 19.58 W
				H = 14 02 30.2 h = normal MB = 4.3
				D = 22.36 Az = 119.9 (NEIS)

Day	Phase		h m s	Remarks
29.	eP	A	14 57 08.5	Kurile Islands 44.39 N 149.67 E H = 14 45 11.3 h = 52 km MB = 4.8 D = 78.58 Az = 334.3 (NEIS) PV A 1.2s 16.3nm M = 4.9
29.	LmV	B	16 50.2	Philippine Islands Region 19.06 N 121.23 E H = 15 53 08.9 h = 59 km MB = 4.7 D = 87.2 LmV B 19s 0.4 μm
29.	ePP	A	21 13 51	Albania 41.54 N 19.03 E
	e	A	15 15	H = 21 11 11.6 h = normal MB = 3.2
	eS	A	15 45	D = 10.46 Az = 333.1 (NEIS)
30.	eP	A	01 16 17	Southern Italy 41.73 N 13.87 E H = 01 14 06.0 h = normal MB = 4.4 D = 9.06 Az = 350.9 (NEIS) PV A 1.0s 15.7nm M = 5.1
30.	eP	A	02 56 04.5	Burma 20.93 N 93.22 E H = 02 45 00.6 h = 32 km MB = 4.9 D = 68.85 Az = 317.5 (NEIS)
30.	eP	A	03 08 25	North of Svalbard 81.50 N 5.14 W
	eX	A	08 31	H = 03 02 05.5 h = normal MB = 4.6 D = 31.43 Az = 159.4 (NEIS) XV A 1.7s 45.5nm
30.	ePKP2	A	04 52 33	Kermadec Islands 29.20 S 177.46 W H = 04 32 12.6 h = 63 km MB = 4.8 (NEIS) D = 157.51
31.	e	A	18 13 21	Near South Coast of France
	e	A	14 04.5	43.40 N 5.57 E
	eSg	A	14 50	H = 18 10 08.7 h = 33 km (NEIS) D = 8.34

Day	Phase		h m s	Remarks
31.	eP	A	23 18 58.5	Hokkaido, Japan Region 42.82 N 145.31 E H = 23 07 01.6 h = 54 km MB = 5.1 D = 78.55 Az = 332.0 (NEIS) PV A 1.2s 20.3nm M = 5.0

Day	Phase		h m s	Remarks
1.	ePKIKP	A	06 58 41	South of Fiji Islands 24.16 S 176.35 W
	ePKHKP	A	58 48.5	H = 06 38 54.1 h = 35 km MB=5.7 MS=5.3
	ePKP2	A	58 59	D = 152.85 Az = 348.9 (NEIS)
	ePP	A	07 02 31	PKIKPV A 1.5s 25.1nm
	e	A	02 38.5	PKHKPV A 1.2 71.1nm
	LmH	B	08 17.8	PKP2V A 1.3 96.1nm
	LmV	B	20.6	LmH B 18 0.4/um
				LmV B 17 0.5/um
1.	ePn	A	11 24 05	Yugoslavia 45.94 N 15.9 E
	eSn	A	25 12	H = 11 22 41 h = 0 km
				D = 5.51 Az = 330 (ISC)
1.	eP	A	18 23 23	Southern Greece 37.2 N 21.0 E
				H = 18 19 46 h = 0 km
				D = 15.05 Az = 336 (ISC)
2.	eP	A	05 52 56	Western Iran 32.66 N 48.21 E
	eS	B	57 46	H = 05 46 37.6 h = 55 km MB = 4.9
	LmH	B	06 09.9	D = 32.29 Az = 314.7 (NEIS)
	LmV	B	09.4	LmH B 16s 2.0/um M = 4.8
				LmV B 13.5 1.8/um 4.8
2.	eP	A	06 03 56.5	Western Iran 32.65 N 48.24 E
				H = 05 57 33.0 h = 59 km MB = 4.9
				D = 32.32 Az = 314.7 (NEIS)
2.	eP1	A	07 41 52	Eastern Russia 54.12 N 125.84 E
	eP2	A	41 57	H = 07 31 33.5 h = normal MB=4.9 MS=5.1
	LmH	B	08 10.4	D = 61.99 Az = 318.9 (NEIS)
	LmV	B	13.6	P1V A 1.6s 19.2nm M = 5.0
				P2V A 1.6 44.0nm 5.3
				LmH B 13 3.2/um 5.7
				LmV B 12 1.8/um 5.5
2.	eP	A	12 20 31.5	India-East Pakistan Border Region
	e	A	22 27	25.72 N 91.60 E
				H = 12 09 55.2 h = 20 km MB = 4.8
				D = 64.36 Az = 315.9 (NEIS)
				PV A 1.2s 12.2nm M = 4.9

Day	Phase		h m s	Remarks
3.	+iP	A	00 31 11	Near East Coast of Kamchatka 54.62 N 161.45 E H = 00 19 51.5 h = 61 km MB = 5.3 D = 72.11 Az = 340.4 (NEIS)
3.	eP	A	01 58 30.5	Northern Colombia 7.26 N 74.35 W H = 01 46 05.3 h = 18 km MB = 5.0 D = 81.88 Az = 39.9 (NEIS)
3.	eP	A	06 05 31.5	Unimak Island Region 53.56 N 163.24 W H = 05 53 45.6 h = 40 km MB = 4.1 D = 76.08 Az = 3.4 (NEIS)
3.	eP	A	08 39 09.5	Carlsberg Ridge 9.87 N 57.89 E
	LmH	B	09 09.2	H = 08 29 35.3 h = normal MB=5.1 MS=5.4
	LmV	B	09.2	D = 55.59 Az = 326.1 (NEIS)
				PV A 2.4s 69.0nm M = 5.3
3.	eP	A	14 31 39	Catamarca Province, Argentina
	ePP	BC	35 52	26.08 S 67.77 W
	eSKS	BC	42 24	H = 14 17 41.8 h = 36 km MB=5.9 MS=5.8
	ePS	BC	45 00	D = 103.39 Az = 40.0 (NEIS)
	eSS	BC	50 20	LmH B 18.2s 2.7/um M = 5.8
	LmH	B	15 23.6	LmV B 17 2.8/um 5.9
	LmV	B	23.5	
4.	ePKP	A	12 19 33.5	Banda Sea 6.03 S 124.69 E H = 12 02 09.1 h = 605 km MB = 5.5 D = 109.18 Az = 321.7 (NEIS)
4.	eP	A	13 00 25.5	Southern Yukon Territory, Canada 61.57 N 140.44 W H = 12 49 37.6 h = 7 km MB = 4.6 D = 65.92 Az = 19.1 (NEIS)
4.	eP	A	13 12 41	Sakhalin Island 53.87 N 141.42 E H = 13 01 45.2 h = normal MB = 5.1 D = 67.67 Az = 328.1 (NEIS) PV A 1.4s 18.6nm M = 5.0

Day	Phase		h m s	Remarks
4.	-iP	ABC	15 55 21.8	Greece 38.90 N 20.44 E
	LmH	B	16 00.2	H = 15 52 11.7 h = 8 km MB=5.8 MS=5.5
	LmV	B	01.5	D = 13.30 Az = 334.9 (NEIS)
				PV A 0.9s 661.0nm M = 6.7
				LmH B 15.5 60.9 µm 5.8
				LmV B 10 28.6 µm
4.	eP	A	16 14 45.5	Greece - Albania Border Region
				39.08 N 20.46 E
				H = 16 11 36.0 h = 6 km MB = 5.0
				D = 13.15 Az = 334.5 (NEIS)
				PV A 0.9s 42.8nm M = 5.6
4.	eP	A	20 10 33	Off East Coast of Kamchatka
	LmH	B	45.3	52.46 N 160.82 E
	LmV	B	50.0	H = 19 58 59.0 h = normal MB = 4.8
				D = 74.02 Az = 340.2 (NEIS)
5.	eP	A	02 28 57	Guerrero, Mexico 18.68 N 100.59 W
				H = 02 16 08.5 h = 88 km MB = 5.1
				D = 89.00 Az = 36.1 (NEIS)
5.	eP	A	08 42 58	Southern Italy 41.68 N 13.76 E
	LmH	B	47.6	H = 08 40 46.7 h = normal MB = 4.0
	LmV	B	48.3	D = 9.09 Az = 351.3 (NEIS)
				LmH B 14s 1.4 µm M = 3.9
				LmV B 10 1.2 µm
5.	ePKHKP	A	11 38 52	South of Fiji Islands 24.29 S 179.17 E
	ePKP2	A	39 03.5	H = 11 19 59.3 h = 580 km MB = 5.5
				D = 151.97 Az = 343.0 (NEIS)
5.	ePKP2	A	14 30 18.5	Fiji Islands Region 21.09 S 178.16 W
				H = 14 11 36.3 h = 629 km MB = 4.6
				D = 149.52 Az = 347.7 (NEIS)
				PKP2V A 1.0s 11.8nm

Day	Phase		h m s	Remarks
5.	e	A	20 17 29.5	Turkey 37.94 N 42.81 E
	e	A	17 35	H = 20 11 57.8 h = 47 km MB = 4.7
	LmH	B	28.9	D = 25.48 Az = 309.9 (NEIS)
	LmV	B	29.0	LmH B 20s 1.1 µm M = 4.4
				LmV B 18 0.7 µm 4.3
6.	ePKIKP	A	05 38 04	South of Fiji Islands 23.79 S 179.14 E
	ePKHKP	A	38 11.5	H = 05 19 18.3 h = 546 km MB = 5.5
	ePKP2	A	38 23	D = 151.50 Az = 343.2 (NEIS)
				PKIKPV A 2.0s 38.5nm
				PKHKPV A 2.0 111.0nm
				PKP2V A 1.5 80.4nm
6.	ePKHKP	A	06 22,56.5	South of Fiji Islands 23.86 S 179.04 E
				H = 06 04 10.2 h = 594 km MB = 4.8
				D = 151.53 Az = 343.1 (NEIS)
6.	eP	A	09 31 19	South Atlantic Ridge 15.59 S 13.64 W
				H = 09 20 12.2 h = normal MB = 4.9
				D = 69.58 Az = 16.9 (NEIS)
				PV A 1.4s 20.9nm M = 5.0
6.	eP1	ABC	09 48 00.5	Andreanof Islands, Aleutian Is.
	eP2	A	48 06	51.62 N 175.40 W
	LmH	B	10 28.4	H = 09 36 05.0 h = 34 km MB=5.8 MS=6.4
	LmV	B	28.2	D = 77.94 Az = 355.4 (NEIS)
				P1V A 1.5s 60.3nm M = 5.4
				P1V B 18 4.0 µm 6.1
				P2V A 1.1 60.5nm 5.5
				LmH B 18.5 12.5 µm 6.3
				LmV B 18 8.5 µm 6.2
6.	eP	ABC	18 38 30	Andreanof Islands, Aleutian Is.
	eS	BC	48 22	51.58 N 175.25 W
	eSS	BC	53 28	H = 18 26 35.1 h = 41 km MB=5.9 MS=6.3
	eSSS	B	58 48	D = 77.99 Az = 355.5 (NEIS)
	LmH	B	19 18.5	PV A 1.7s 127.0nm M = 5.7
	LmV	B	21.6	PV B 11 3.5 µm 6.3
				LmH B 19 12.4 µm 6.3
				LmV B 16.5 9.5 µm 6.2

Day	Phase		h m s	Remarks
6.	eP	A	20 42 04.5	Greece 38.13 N 20.26 E
				H = 20 38 46.5 h = 14 km MB = 4.2
				D = 13.94 Az = 336.6 (NEIS)
6.	e(PKP)	A	23 22 57.5	Tonga 19.7 S 174.20 W
				H = 23 03 13.2 h = 0 km MB = 4.7
				D = 148.76 Az = 353 (ISC)
				(PKP)V A 2.0s 25.6nm
7.	ePKP	A	05 48 09	Tonga Islands 15.25 S 173.55 W
				H = 05 28 35.1 h = normal MB = 4.8
				D = 144.45 Az = 354.3 (NEIS)
				PKPV A 1.2s 10.2nm
7.	ePKIKP	A	07 37 37	New Ireland Region 5.93 S 153.56 E
	e	A	37 52.5	H = 07 18 39.8 h = normal MB=5.1 MS=5.0
	e	A	39 43.5	D = 125.30 Az = 331.3 (NEIS)
	LmH	B	08 33.4	PV A 2.4s 69.0nm
	LmV	B	36.1	LmH B 15 0.8/um
				LmV B 18 1.3/um
7.	eX	A	15 09 51.5	Hokkaido, Japan Region
				43.42 N 145.33 E
				H = 14 57 13.5 h = 76 km MB = 4.7 (NEIS)
				D = 78.02
				XV A 1.9s 22.7nm
7.	eP	A	17 07 54	Northern Italy 44.00 N 11.23 E
				H = 17 06 15.5 h = normal MB = 4.2
				D = 6.66 Az = 2.1 (NEIS)
7.	ePKHKP	A	21 17 40	South of Fiji Islands 22.34 S 176.85 W
				H = 20 58 09.5 h = 192 km MB = 4.9
				D = 150.98 Az = 348.9 (NEIS)
8.	eP	ABC	09 10 50.5	Kurile Islands 50.05 N 156.31 E
	eS	BC	20 24	H = 08 59 10.0 h = normal MB=6.0 MS=6.1
	eSS	C	25 32	D = 75.25 Az = 337.6 (NEIS)

Day	Phase		h m s	Remarks
ont. 8.	eSSS	C	09 28 44	PV A 2.1s 412.0nm M = 6.1
	LmH	B	48.5	PV B 8 2.7/um 6.4
	LmV	B	48.4	LmH B 18 19.7/um 6.4
				LmV B 18 16.9/um 6.4
8.	eP	A	13 31 01	Guerrero, Mexico 18.13 N 99.69 W
				H = 13 18 12.0 h = 71 km MB = 5.3
				D = 88.93 Az = 36.4 (NEIS)
8.	eP	A	15 54 50	Jordan - Syria Region 35.5 N 36.2 E
				H = 15 49 49 h = 33 km
				D = 23.34 Az = 318 (ISC)
				PV A 1.8s 20.3nm M = 4.3
9.	e(P)	A	07 10 11.5	Mediterranean Sea 35.78 N 21.99 E
				H = 07 06 24.0 h = normal MB = 4.2
				D = 16.65 Az = 336.4 (NEIS)
9.	ePKP	A	08 28 01	Samoa Islands Region 16.12 S 172.59 W
				H = 08 08 25.0 h = normal MB=5.1 MS=4.6
				D = 145.39 Az = 355.3 (NEIS)
				PV A 1.6s 38.5nm
9.	eP1	AB	13 49 43	North of Svalbard 86.06 N 32.68 E
	eP2	A	49 48	H = 13 42 43.7 h = normal MB=5.3 MS=5.1
	eS	BC	55 21	D = 35.86 Az = 203.0 (NEIS)
	LmH	B	14 12.9	P1V A 2.0s 145.0nm M = 5.5
	LmV	B	12.8	P2V A 2.0 222.0nm 5.6
				LmH B 16 4.3/um 5.3
				LmV B 16.5 5.8/um 5.5
9.	eP	A	14 24 35	Rat Islands, Aleutian Is.
				52.44 N 178.35 E
				H = 14 13 03.5 h = 182 km MB = 5.6
				D = 76.68 Az = 351.4 (NEIS)
				PV A 1.6s 33.0nm M = 4.8

Day	Phase		h m s	Remarks
9.	eP	A	14 54 36	North of Svalbard 86.00 N 33.23 E H = 14 47 38.1 h = normal MB=5.0 MS=4.9 D = 35.83 Az = 203.6 (NEIS) PV -A 1.5s 55.2 M = 5.3
9.	eP	A	15 16 33.5	North of Svalbard 85.98 N 31.26 E H = 15 09 36.1 h = normal MB=5.2 MS=5.0 D = 35.75 Az = 201.5 (NEIS) PV A 2.5s 200.0nm M = 5.6
9.	e	A	18 58 29	Southern Iran 27.72 N 55.52 E H = 18 50 50.4 h = 39 km MB = 4.5 D = 40.23 Az = 316.9 (NEIS)
9.	eP	A	22 54 47.5	Near Coast of Northern Peru 3.97 S 80.97 W H = 22 41 27.7 h = 31 km MB = 5.6 D = 94.68 Az = 39.7 (NEIS) PV A 1.6s 46.7nm M = 5.7
9.	eP	A	23 38 49	Nicobar Islands Region 6.02 N 93.98 E H = 23 26 39.0 h = normal MB = 5.1 D = 80.53 Az = 320.2 (NEIS) PV A 1.4s 23.2nm M = 5.0
10.	e e	A A	02 35 19 36 18	Poland Rockburst (CLL)
10.	ePn e	A A	03 02 50 04 50	Adriatic Sea 44.02 N 13.23 E H = 03 01 11.6 h = normal MB = 4.4 D = 6.72 Az = 351.2 (NEIS) PV A 1.0s 2.36nm M = 5.0
10.	eP	A	05 17(23)	Hokkaido, Japan Region 41.51 N 142.07 E H = 05 05 31.1 h = 66 km MB = 4.6 D = 78.53 Az = 330.4 (NEIS)

Day	Phase		h m s	Remarks
10.	ePKP	A	09 40 51	New Hebrides Islands Region
				19.07 S 167.41 E
				H = 09 21 16.6 h = 12 km MB = 5.4
				D = 143.09 Az = 334.2 (NEIS)
11.	eP	ABC	02 54 45	Kurile Islands 49.98 N 156.36 E
	eS	C	03 04 26	H = 02 43 06.2 h = 51 km MB = 6.0
	LmH	B	32.3	D = 75.33 Az = 337.7 (NEIS)
	LmV	B	32.3	PV A 1.9s 299.0nm M = 5.9
				LmH B 19 6.8 μm 6.0
				LmV B 18.5 6.4 μm 6.0
11.	eP	A	07 21 59.5	Iran 30.57 N 52.89 E
	eS	C	27 38	H = 07 14 51.5 h = 11 km MB = 5.5
	LmH	B	40.4	D = 36.58 Az = 315.2 (NEIS)
	LmV	B	40.4	PV A 1.4s 25.6nm M = 4.8
				LmH B 15.5 3.5 μm 5.2
				LmV B 16 4.3 μm 5.4
11.	ePKP	A	10 21 32.5	Fiji Islands Region 18.13 S 178.36 W
	e	A	21 35	H = 10 02 56.7 h = 583 km MB = 4.6
				D = 146.59 Az = 348.4 (NEIS)
11.	e	A	11 39 33	eV A 1.2s 16.3nm
11.	ePKHKP	A	17 12 59.5	South of Fiji Islands 22.10 S 179.53 W
	epPKIKP	A	15 17.5	H = 16 54 11.6 h = 568 km MB = 5.1
				D = 150.20 Az = 345.7 (NEIS)
				PKHKPV A 1.0s 23.6nm
				pPKIKPV A 1.4s 11.6nm
11.	e	A	20 07 27	
11.	ePKIKP	A	22 14 26	Fiji Islands Region 19.90 S 176.48 W
	ePKHKP	A	14 30	H = 21 55 13.1 h = 274 km MB = 5.4
	ePKP2	A	14 34	D = 148.65 Az = 350.1 (NEIS)
				PKIKPV A 1.5s 50.2nm
				PKHKPV A 1.6 330.0nm
				PKP2V A 1.2 77.2nm

Day	Phase		h m s	Remarks
12.	eP	AB	00 11 34.5	Dodecanese Islands 35.49 N 27.66 E H = 00 07 13.3 h = 69 km MB = 4.7 D = 19.09 Az = 327.4 (NEIS) PV A 1.1s 40.3nm M = 4.6
12.	eP	ABC	00 16 09	Dodecanese Islands 35.58 N 27.69 E
	e	A	16 15.5	H = 00 11 52.3 h = normal MB = 5.1
	LmH	B	23.9	D = 19.03 Az = 327.3 (NEIS)
	LmV	B	26.6	P1V A 1.2s 24.4nm M = 4.3 eV A 1.3 120.0nm 5.0 LmH B 14.5 8.4/um 5.2 LmV B 11 7.8/um 5.4
12.	ePKIKP	A	04 12 43	Solomon Islands 6.15 S 154.46 E H = 03 53 44.0 h = 50 km MB=5.6 MS=5.9 D = 125.92 Az = 331.7 (NEIS)
12.	e(P)	A	04 21 02.5	Dodecanese Islands 35.33 N 27.93 E H = 04 16 39.0 h = 71 km D = 19.34 Az = 327 (ISC)
12.	eP	A	07 58 31.5	Dodecanese Islands 35.29 N 27.30 E H = 07 54 07.5 h = normal MB = 4.1 D = 19.10 Az = 328.3 (NEIS)
13.	eP	A	01 24 18	Near East Coast of Honshu, Japan 38.62 N 142.15 E H = 01 12 11.5 h = 78 km MB = 5.5 D = 81.07 Az = 330.7 (NEIS) PV A 1.4s 30.2nm M = 5.0
13.	eP	A	02 58 18	Northwest of Kurile Islands 49.60 N 151.21 E H = 02 47 14.9 h = 326 km MB = 5.1 D = 74.32 Az = 334.6 (NEIS) PV A 1.7s 42.5nm M = 4.9

Day	Phase		h m s	Remarks
13.	ePKP	A	16 29 36	Fiji Islands Region 18.28 S 178.13 W
	ePKP2	A	29 38.5	H = 16 10 58.9 h = 571 km MB = 5.6
				D = 146.78 Az = 348.7 (NEIS)
				PKPV A 1.6s 60.4nm
				PKP2V A 1.3 205.2nm
13.	ePKP	A	22 50 04.5	Fiji Islands Region 17.74 S 178.73 W
	e	A	50 06.5	H = 22 31 26.3 h = 554 km MB = 5.1
				D = 146.14 Az = 348.2 (NEIS)
13.	eP	A	22 58 41	Western Iran 32.70 N 48.31 E
				H = 22 52 15.1 h = 59 km MB = 4.9
				D = 32.33 Az = 314.6 (NEIS)
				PV A 2.0s 38.5nm M = 4.9
14.	e	A	01 17(23)	
	e	A	17 30	
14.	ePn	A	06 52 19	Austria 47.6 N 14.1 E
				H = 06 51 20 h = 0 km
				D = 3.50 Az = 333 (ISC)
14.	eP	A	09 38 19.5	Dodecanese Islands 35.44 N 27.68 E
	e	A	38 22.5	H = 09 33 58.5 h = 55 km MB = 4.4
	LmH	B	45.6	D = 19.14 Az = 327.5 (NEIS)
	LmV	B	48.3	PV A 1.4s 23.2nm M = 4.2
				eV A 1.7 45.5nm 4.4
				LmH B 12.5 1.0 μm 4.3
				LmV B 14 1.0 μm 4.4
14.	e(P)	A	15 57 47.5	Southern Sinkiang Prov., China
				40.73 N 78.75 E
				H = 15 49 16.7 h = 35 km MB = 4.7
				D = 46.43 Az = 305.9 (NEIS)
15.	LmH	B	03 14.0	Atlantic Ocean 56.9 N 34.4 W
	LmV	B	15.0	H = 02 57 22 h = 0 km MB = 4.5 (ISC)
				D = 27.5

Day	Phase		h m s	Remarks
15.	eP1	AB	06 15 42	<u>North of Ascension Island</u>
	eP2	A	15 48	1.36 S 15.94 W
	eS	BC	23 40	H = 06 05 57.4 h = normal MB=5.2 MS=4.9
	LmH	B	39.6	D = 56.90 Az = 20.6 (NEIS)
	LmV	B	43.0	P1V A 2.5s 138.0nm M = 5.5
				P2V A 2.0 137.0nm 5.6
				LmH B 16.5 1.4 µm 5.1
				LmV B 14.5 1.2 µm 5.2
15.	eP	A	08 32 03	<u>Nicobar Islands Region</u> 9.91 N 93.75 E
				H = 08 20 09.3 h = normal MB = 5.2
				D = 77.43 Az = 319.7 (NEIS)
15.	ePg	A	08 35 37	<u>Austria</u> 47.5 N 14.0 E
				H = 08 34 38 h = 0 km
				D = 3.56 Az = 335 (ISC)
15.	eP	AB	15 16 19	<u>North of Ascension Island</u>
	e	A	16 26	1.37 S 15.79 W
	eS	BC	24 16	H = 15 06 35.5 h = normal MB=5.5 MS=5.0
	LmH	B	43.6	D = 56.86 Az = 20.5 (NEIS)
	LmV	B	43.7	PV A 2.6s 225.0nm M = 5.7
				LmH B 16 1.5 µm 5.2
				LmV B 14 1.4 µm 5.3
16.	ePKP	A	18 22 31	<u>Fiji Islands Region</u> 18.13 S 178.39 W
				H = 18 03 55.4 h = 608 km MB = 4.7 (NEIS)
				D = 146.58
				PKPV A 1.1s 20.2nm
16.	ePKIKP	A	19 50 57.5	<u>Tonga Islands</u> 20.59 S 174.27 W
				H = 19 31 16.1 h = normal
				MB = 4.7 MS = 4.9 (NEIS)
				D = 149.53 Az = 353 (ISC)
16.	ePKHKP	A	23 28 33	<u>South of Austria</u> 51.18 S 139.53 E
	LmH	B	00 39.7	H = 23 08 49.0 h = normal

Day	Phase		h m s	Remarks
cont. 16.	LmV	B	00 39.8	MB = 5.4 MS = 5.8 (NEIS)
				D = 147.72
				PKHKPV A 1.7s 48.5nm
				LmH B 20.5 1.4 µm M = 5.7
				LmV B 20 1.6 µm
17.	eP	A	00 38 32	Kurile Islands 43.15 N 147.33 E
				H = 00 26 27.3 h = normal MB = 5.0 (NEIS)
				D = 78.93
				PV A 1.6s 19.2nm M = 4.9
17.	eP1	AB	11 02 43	Carlsberg Ridge 1.60 S 69.83 E
	eP2	ABC	02 48	H = 10 51 21.8 h = normal
	eS	BC	11 54	MB = 5.5 MS = 5.5 (NEIS)
	eSS	BC	16 36	D = 71.71
	LmH	B	36.8	P1V A 2.4s 152.0nm M = 5.7
	LmV	B	36.9	P2V A 2.4 276.0nm 6.0
				LmH B 18 1.1 µm 5.2
				LmV B 17 1.0 µm 5.2
18.	eP	A	12 19 32.5	Kurile Islands 44.46 N 148.78 E
				H = 12 07 35.8 h = 50 km (NEIS)
				D = 78.23
				PV A 1.7s 48.5nm M = 5.2
19.	eP	A	07 33 21	Dodecanese Islands 35.37 N 27.58 E
	LmH	B	41.0	H = 07 28 57.6 h = normal MB = 4.7
	LmV	B	43.3	D = 19.16 Az = 327.7 (NEIC)
				PV A 1.7s 63.6nm M = 4.6
				LmH B 14 1.6 µm 4.5
				LmV B 14 1.3 µm 4.5
19.	eP	A	07 38 13.5	Dodecanese Islands 35.35 N 27.86 E
				H = 07 33 51.5 h = 83 km
				D = 19.30 Az = 310 (ISC)
				PV A 1.3s 17.5nm M = 4.2

Day	Phase		h m s	Remarks
19.	eP	A	11 33 20.5	Salta Province, Argentina
	LmH	B	12 21.7	24.68 S 64.60 W
	LmV	B	17.8	H = 11 19 35.1 h = 40 km MB=5.8 MS=5.9
				D = 100.49 Az = 39.0 (NEIS)
				LmH B 17s 2.8 µm M = 5.8
				LmV B 19 2.8 µm 5.8
19.	+iP	AB	13 14 04.5	Near East Coast of Honshu, Japan
	ePP	B	17 08	38.91 N 141.91 E
	eSKS	BC	24 10	H = 13 01 56.1 h = 56 km MB = 6.1
	eSS	B	29 26	D = 80.73 Az = 330.5 (NEIS)
	eSSS	B	33 00	PV A 1.2s 382.0nm M = 6.3
	LmH	B	51.4	PV B 12 4.2 µm 6.3
	LmV	B	53.2	LmH B 18.7 28.2 µm 6.6
				LmV B 16 21.1 µm 6.6
19.	ePKP	A	18 58 25	Central Chile 28.46 S 70.85 W
	LmH	B	19 42.7	H = 18 40 05.2 h = 58 km MB = 5.7
	LmV	B	49.6	D = 106.99 Az = 41.3 (NEIS)
				LmH B 19s 2.1 µm M = 5.7
				LmV B 18 2.0 µm 5.7
19.	eP	A	21 23 25	Near East Coast of Honshu, Japan
	epP	A	23 39	38.89 N 141.92 E
	LmH	B	22 00.5	H = 21 11 17.1 h = 60 km MB = 5.1
	LmV	B	02.7	D = 80.75 Az = 330.5 (NEIS)
				PV A 1.6s 52.2nm M = 5.2
				LmH B 18 0.9 µm 5.1
				LmV B 16 1.2 µm 5.4
20.	ePKP	A	00 44 53	South of Tonga 25.74 S 176.0 W
	e	A	45 58	H = 00 24 49.5 h = 33 km
				D = 154.45 Az = 349 (ISC)
20.	ePKP	A	11 47 02.5	Tonga 15.43 S 174.77 W
				H = 11 27 57.3 h = 262 km MB = 4.8
				D = 144.49 Az = 353 (ISC)

Day	Phase		h m s	Remarks
20.	eP	A	13 06 03	**Aegean Sea** 39.42 N 23.96 E
	eX	A	06 06.5	H = 13 02 36.8 h = normal MB = 4.8
	LmH	B	11.7	D = 14.19 Az = 326.3 (NEIS)
	LmV	B	13.8	PV A 1.2s 26.4nm M = 4.8
				XV A 2.0 59.9nm 5.0
				LmH B 13.2 6.0 μm 4.9
				LmV B 13 2.9 μm
20.	eP	A	17 09 13	**Off East Coast of Kamchatka**
	LmH	B	44.0	52.52 N 160.88 E
	LmV	B	48.1	H = 16 57 38.6 h = normal MB=5.0 MS=4.9
				D = 73.97 Az = 340.2 (NEIS)
				PV A 1.1s 24.2nm M = 5.1
				LmH B 16 1.8 μm 5.5
				LmV B 13.5 1.8 μm 5.6
20.	eP	A	19 23 46.5	**Southern Sumatra** 1.35 S 100.63 E
	e	A	23 51	H = 19 10 45.6 h = normal MB = 5.0
				D = 90.40 Az = 320.5 (NEIS)
21.	e(P)	A	01 00 12	**Lake Tanganyika Region** 3.61 S 28.19 E
				H = 00 50 30.2 h = normal MB = 5.1
				D = 55.89 Az = 347.3 (NEIS)
				traces
21.	eP	ABC	13 44 17.5	**Near Coast of Chiapas, Mexico**
	e(S)	BC	54 56	14.50 N 92.73 W
	LmH	B	14 25.5	H = 13 31 32.1 h = 59 km MB = 5.3
	LmV	B	25.6	D = 87.76 Az = 38.1 (NEIS)
				PV A 1.4s 14.0nm M = 5.0
				LmH B 18 2.8 μm 5.7
				LmV B 17.5 2.8 μm 5.8
21.	ePKIKP	A	15 20 27	**Fiji Islands Region** 20.07 S 178.42 W
	ePKHKP	A	20 31.5	H = 15 01 52.6 h = 617 km MB = 5.1
	ePKP2	A	20 37.5	D = 148.47 Az = 347.8 (NEIS)
				PKIKPV A 1.7s 30.3nm
				PKHKPV A 1.6 184.0nm
				PKP2V A 1.5 77.9nm

Day	Phase		h m s	Remarks
21.	eP1	A	19 57 03	Tibet 34.66 N 80.97 E
	eP2	A	57 05.5	H = 19 47 56.3 h = 26 km MB=5.0 MS=5.2
	LmH	B	20 17.6	D = 51.57 Az = 310.5 (NEIS)
	LmV	B	21.4	P1V A 1.2s 10.2nm M = 4.6
				P2V A 1.1 22.2nm 5.0
				LmH B 19 2.1 μm 5.2
				LmV B 14 1.4 μm 5.2
21.	+iP	A	21 17 08	Kurile Islands 46.13 N 151.45 E
				H = 21 05 20.4 h = 84 km MB = 5.6
				D = 77.54 Az = 335.1 (NEIS)
				PV A 1.4s 181.0nm M = 5.8
22.	eP	A	01 40 04	Southern Yukon Territory, Canada
				61.35 N 138.98 W
				H = 01 29 20.4 h = normal MB = 4.5
				D = 65.90 Az = 20.0 (NEIS)
22.	eP	A	08 00 15	Crete 35 N 27 E
				H = 07 55 50 (NORSAR)
				D = 19.3
22.	e(P)	A	14 58 39.5	Turkey 40.31 N 29.81 E
				H = 14 54 55.9 h = normal MB = 4.2
				D = 16.38 Az = 315.2 (NEIS)
				traces
22.	eP	A	18 07 58.5	North Atlantic Ridge 22.6 N 45.2 W
				H = 17 58 20 h = 0 km MB = 4.1
				D = 51.94 Az = 43 (ISC)
23.	e(PKP)	A	10 15 09.5	South of Tonga 24.2 S 175.7 W
				H = 09 55 14 h = 0 km
				D = 153.01 Az = 350 (ISC)
23.	eP	A	13 42 34.5	Azores Islands 38.46 N 28.31 W
	e	ABC	42 41.5	H = 13 36 19.3 h = 5 km MB=5.0 MS=5.1
	ePP	BC	43 28	D = 30.58 Az = 53.4 (NEIS)

Day	Phase		h m s	Remarks
cont. 23.	eS	BC	13 47 39	PV A 1.3s 15.3nm M = 4.8
	eSS	BC	49 31	LmH B 12.8 4.8 um 5.3
	LmH	B	53.8	LmV B 15 4.9 um 5.4
	LmV	B	53.9	
23.	ePKP	A	20 15 31	Tonga Islands 16.63 S 174.57 W
				H = 19 56 08.5 h = 150 km MB = 4.5
				D = 145.71 Az = 353.0 (NEIS)
23.	ePKP	A	20 34 27.5	Off Coast of Southern Chile
				44.97 S 80.28 W
				H = 20 15 30.4 h = normal MB=5.3 MS=5.1
				D = 123.93 Az = 50.1 (NEIS)
24.	ePKHKP	A	11 03 54.5	Tonga Islands 18.22 S 173.38 W
				H = 10 44 13.8 h = 42 km MB = 4.6
				D = 147.40 Az = 354.1 (NEIS)
24.	eP1	A	14 09 25	Algeria 36.14 N 4.41 E
	eP2	A	09 29	H = 14 05 46.4 h = 17 km MB=5.1 MS=5.1
	LmH	B	14.8	D = 15.40 Az = 17.5 (NEIS)
	LmV	B	17.0	P1V A 1.1s 24.2nm M = 4.6
				P2V A 1.3 54.5nm 4.9
				LmH B 16.5 8.2 um 5.0
				LmV B 13 3.5 um 4.9
24.	eP1	A	15 25 47	Algeria 36.12 N 4.44 E
	eP2	A	25 51	H = 15 22 09.8 h = normal MB=5.1 MS=5.3
				D = 15.42 Az = 17.4 (NEIS)
				P1V A 1.5s 30.2nm M = 4.4
				P2V A 2.3 195.0nm 5.0
25.	eP1	A	04 23 59	Algeria 36.19 N 4.47 E
	eP2	A	24 02.5	H = 04 20 22.5 h = normal MB=4.9 MS=4.4
	LmH	B	31.8	D = 15.34 Az = 17.4 (NEIS)
	LmV	B	33.5	P1V A 1.0s 15.8nm M = 4.5
				P2V A 1.2 46.7nm 4.8
				LmH B 10 2.2 um 4.6
				LmV B 8 1.7 um 4.8

Day	Phase		h m s	Remarks
25.	eP	A	04 37 05.5	Near South Coast of Southern Honshu
	e	A	37 20	33.82 N 135.45 E
				H = 04 24 47.3 h = 51 km MB = 5.7
				D = 82.41 Az = 327.8 (NEIS)
25.	eP	A	04 51 33	Mediterranean Sea 35.1 N 17.7 E
				H = 04 47 41 h = 0 km
				D = 16.19 Az = 346 (ISC)
25.	eP	A	07 43 42	North of Ascension Islands
				3.01 S 12.06 W
				H = 07 33 52.5 h = normal MB=4.8 MS=4.6
				D = 57.18 Az = 17.7 (NEIS)
25.	eP	AB	09 31 32.5	Near South Coast of Southern Honshu
	e	A	33 44	33.82 N 135.38 E
	LmH	B	10 11.8	H = 09 19 14.6 h = 56 km MB = 5.7
	LmV	B	12.0	D = 82.38 Az = 327.7 (NEIS)
				PV A 1.7s 97.0nm M = 5.5
				LmH B 16 0.9 μm 5.2
				LmV B 16 0.6 μm 5.1
25.	ePKHKP	A	12 05 36	Tonga Islands 21.45 S 174.17 W
				H = 11 45 49.9 h = normal MB=4.9 MS=4.8
				D = 150.51 Az = 352.5 (NEIS)
25.	e(P)	A	15 37 01	Algeria 36.10 N 4.3 E
				H = 15 33 11 h = 36 km MB = 4.0
				D = 15.48 Az = 18 (ISC)
26.	e(PKIKP)	A	05 43 09	D'Entrecasteaux Islands Region
				8.58 S 154.01 E
				H = 05 23 56.4 h = 15 km MB=5.4 MS=4.8
				D = 127.81 Az = 330.5 (NEIS)
				(PKIKP)V A 1.3s 10.9nm
26.	eP	A	15 49 41.5	Hokkaido, Japan Region 41.58 N 142.04 E
				H = 15 37 48.1 h = 73 km MB = 4.6
				D = 78.46 Az = 330.4 (NEIS)

Day	Phase		h m s	Remarks
26.	e	A	19 04 30	
26.	e(PKP)	A	21 49 50.5	Loyalty Islands Region 21.98 S 169.55 E H = 21 30 05.5 h = normal D = 146.58 Az = 334.3 (NEIS)
27.	eP	A	06 31 27	Kurile Islands 42.96 N 146.28 E H = 06 19 27.7 h = 57 km MB = 4.7 D = 78.75 Az = 332.5 (NEIS)
27.	ePKP2	A	08 21 57.5	South of Fiji Islands 24.82 S 179.76 W H = 08 02 39.5 h = 462 km MB = 4.8 D = 152.75 Az = 344.1 (NEIS) traces
27.	eP	A	09 40 50.5	Tibet 33.01 N 86.72 E H = 09 31 10.8 h = normal MB = 5.0 D = 56.25 Az = 312.3 (NEIS) PV A 1.6s 22.0nm M = 5.0
27.	+iP	AB	14 03 54	Near East Coast of Kamchatka
	eS	BC	13 16	53.65 N 160.53 E
	eSS	C	17 48	H = 13 52 29.6 h = 60 km MB = 5.9
	LmH	B	42.2	D = 72.84 Az = 339.9 (NEIS)
	LmV	B	42.1	PV A 1.1s 565.0nm M = 6.4
				LmH B 16 2.1 µm 5.5
				LmV B 16 2.1 µm 5.5
27.	ePKHKP	A	16 26 48.5	South of Fiji Islands 24.18 S 179.75 W H = 16 07 46.1 h = 460 km MB = 4.8 D = 152.14 Az = 344.4 (NEIS)
28.	ePKHKP	A	01 09 48.5	Fiji Islands Region 20.57 S 177.85 W H = 00 50 49.5 h = 443 km MB = 4.8 D = 149.07 Az = 348.2 (NEIS) PKHKPV A 1.4s 18.6nm

Day	Phase		h m s	Remarks
28.	eP	A	05 35 39.5	<u>Near West Coast of Honshu, Japan</u> 37.79 N 138.30 E H = 05 23 52.0 h = 216 km MB = 4.8 D = 80.26 Az = 328.8 (NEIS)
28.	eP	A	05 52 39	<u>Kurile Islands</u> 43.32 N 147.54 E H = 05 40 41.6 h = 63 km MB = 4.5 D = 78.85 Az = 333.2 (NEIS)
28.	eP	A	06 57 21.5	<u>Guatemala</u> 14.50 N 90.00 W H = 06 45 06.3 h = 230 km MB = 5.0 D = 86.12 Az = 38.7 (NEIS) PV A 1.9s 37.9nm M = 4.9
28.	eP	A	07 18 11	<u>Northern Sumatra</u> 2.87 N 96.36 E H = 07 05 39.0 h = normal MB = 5.2 D = 84.45 Az = 320.4 (NEIS)
28.	eP e LmH LmV	A B B B	08 25 57 44 06 09 14.0 14.6	<u>Prince Edward Islands Region</u> 41.86 S 42.77 E H = 08 12 31.4 h = normal MB=5.7 MS=5.9 D = 96.07 Az = 340.7 (NEIS) LmH B 17s 5.0/um M = 6.1 LmV B 17.5 6.6/um 6.2
28.	eSg	A	23 03 53	<u>Poland</u> 50.18 N 18.51 E H = 23 01 21.0 (ISC) D = 4.45
29.	LmH LmV	B B	01 42.8 42.8	<u>Near North Coast of New Guinea</u> 3.49 S 145.95 E H = 00 29 56.4 h = normal MB = 5.1 MS = 5.7 (NEIS) D = 119.39 LmH B 18s 2.7/um M = 5.9 LmV B 17 3.0/um 6.0

Day	Phase		h m s	Remarks
29.	eP	AB	11 01 49	Crete 35.18 N 23.80 E
	eS	B	05 20	H = 10 57 42.7 h = 26 km MB = 5.7 MS=5.6
	LmH	B	10.3	D = 17.81 Az = 333.9 (NEIS)
	LmV	B	10.2	PV A 1.0s 161.0nm M = 5.1
				LmH B 14 44.4 μm 5.9
				LmV B 14 47.4 μm 6.1
29.	ePg	A	15 01 16.5	Czechoslovakia 50.18 N 13.29 E
	eSg		01(35)	H = 15 00 54 Explosion
29.	ePn	A	16 49 16	Yugoslavia 45.48 N 18.70 E
	eSn	A	50 31	H = 16 47 33.8 h = normal MB = 4.8
				D = 7.02 Az = 320.0 (NEIS)
				PV A 1.2s 16.3nm M = 4.6
29.	eP	A	17 31 32	Hokkaido, Japan Region
				42.86 N 145.44 E
				H = 17 19 34.4 h = 56 km MB = 4.8
				D = 78.56 Az = 332.1 (NEIS)
29.	eP	A	18 09 53	Sea of Okhotsk 53.30 N 153.36 E
	epP	A	11 38	H = 17 59 21.3 h = 491 km MB = 5.2
				D = 71.54 Az = 335.4 (NEIS)
				PV A 1.1s 46.4nm M = 4.9
				pPV A 1.2 30.5nm
29.	e	A	19 47 34	Southern Greece 36.00 N 24.78 E
				H = 19 43 34.7 h = 163 km MB = 3.7
				D = 17.45 Az = 331.1 (NEIS)
				traces
30.	ePKHKP	AB	08 29 01.5	New Hebrides Islands 15.20 S 167.39 E
	ePKIKP	A	29 06	H = 08 09 55.4 h = 124 km MB = 6.0
	eX	A	29 12	D = 139.59 Az = 336.2 (NEIS)
	e	B	29 40	PKHKPV A 1.3s 24.0nm
	e	B	30 04	PKHKPV B 8 5.4 μm
	e	B	32 06	PKIKPV A 1.3 98.4nm
	e	B	32 36	XV A 1.9 269.0nm
	e	B	38 42	LmH B 20 7.8 μm (M = 6.4)

Day	Phase		h m s	Remarks
cont. 30.	LmH	B	09 21.6	LmV B 21s 4.1 /um
	LmV	B	23.9	e B 42 10 e B 50 00
30.	eP	A	18 36 12	<u>Turkey</u> 36.20 N 30.87 E H = 18 31 36.7 h = 24 km MB = 3.1 D = 20.02 Az = 322.2 (NEIS)

Day	Phase		h m s	Remarks
1.	e	A	00 38 16.5	South Sandwich Islands Region
	ePP	A	39 07	59.20 S 25.56 W
	LmV	B	01 22.2	H = 00 19 26.9 h = 32 km MB=5.4 MS=5.2
	LmH	B	22.7	D = 113.58 Az = 24.8 (NEIS)
				LmH B 21s 0.7/um M = 5.2
				LmV B 20 0.6/um 5.2
1.	+eP	A	10 50 50	Kurile Islands 43.01 N 146.08 E
	eS	C	11 00 48	H = 10 38 51.0 h = 36 km MB=5.4 MS=5.3
	LmH	B	21.8	D = 78.64 Az = 332.4 (NEIS)
	LmV	B	29.5	PV A 1.2s 32.5nm M = 5.2
				LmH B 16 1.5/um 5.4
				LmV B 16 1.5/um 5.5
1.	e(P)	A	14 47 29.5	Peru - Boliva Border Region
				17.73 S 69.73 W
				H = 14 34 10.1 h = 142 km MB = 5.2
				D = 98.17 Az = 39.5 (NEIS)
				PV A 1.7s 18.2nm M = 5.3
1.	eP	A	17 19 33	Off Coast of Central Chile
	LmH	B	18 15.9	35.59 S 74.48 W
	LmV	B	19.9	H = 17 00 56.2 h = 35 km MB=5.8 MS=5.9
				D = 114.30 Az = 44.2 (NEIS)
				LmH B 16s 1.3/um M = 5.6
				LmV B 18 1.3/um 5.6
1.	+eP	A	23 28 56	Kurile Islands 43.10 N 146.94 E
				H = 23 16 54.8 h = 31 km MB=5.3 MS=5.9
				D = 78.85 Az = 332.9 (NEIS)
				PV A 1.9s 72.0nm M = 5.4
1.	+iP	AB	23 30 03.2	Kurile Islands 43.16 N 146.87 E
	eS	B	40 00	H = 23 18 03.9 h = 40 km MB=5.8 MS=5.9
	eSS	C	45 20	D = 78.77 Az = 332.8 (NEIS)
	LmH	B	24 08.5	PV A 1.3s 100.5nm M = 5.7
	LmV	B	11.0	PV B 10 2.16/um 6.1
				LmH B 16 12.7/um 6.4
				LmV B 15.5 10.9/um 6.3

Day	Phase		h m s	Remarks
2.	e(P)	A	00 57 43	Kurile Islands 43.24 N 146.9 E
	e	A	57 49	H = 00 45 38 h = 31 km
				D = 78.72 Az = 333 (ISC)
2.	eP	A	01 38 47	Ascension Island Region
				10.52 S 13.10 W
				H = 01 28 11.6 h = normal MB = 4.9
				D = 64.60 Az = 17.1 (NEIS)
				PV A 1.4s 18.6nm M = 5.0
2.	eP	A	10 29 11	Kurile Islands 43.90 N 147.55 E
				H = 10 17 12.7 h = normal MB = 4.7
				D = 78.34 Az = 333.1 (NEIS)
				PV A 1.4s 18.6nm M = 4.9
2.	ePKHKP	A	13 22 28	South of Fiji Islands 24.39 S 179.86 W
	ePKP2	A	22 41	H = 13 03 26.3 h = 464 km MB = 5.0
				D = 152.32 Az = 344.2 (NEIS)
2.	e(pP)	A	15 03 52	Fiji Islands 16.28 S 178.12 E
				H = 14 44 08.5 h = 61 km MB = 4.9
				D = 144.04 Az = 345.3 (NEIS)
				traces
2.	LmH	B	19 24.5	South Sandwich Islands Region
	LmV	B	26.3	55.77 S 28.25 W
				H = 18 19 47.2 h = 33 km MB = 5.2 (ISC)
				D = 110.0
				LmH B 18s 0.4 μm M = 5.1
				LmV B 16 0.6 μm 5.3
2.	+iP	AB	22 21 47.5	Fox Islands, Aleutian Is.
	epP	A	21 58	52.28 N 168.74 W
	e	A	22 06	H = 22 09 54.5 h = 40 km MB=5.6 MS=5.0
	LmH	B	23 07.0	D = 77.45 Az = 359.8 (NEIS)
	LmV	B	09.0	h = 39 km
				PV A 1.3s 166.0nm M = 5.9
				LmH B 16.8 0.9 μm 5.1
				LmV B 16 0.6 μm 5.0

Day	Phase		h m s	Remarks
3.	LmH	B	03 12.8	<u>Near North Coast of West Irian</u>
	LmV	B	14.0	2.81 S 139.10 E
				H = 02 01 32.8 h = 38 km MB = 5.5 (ISC)
				D = 115.2
				LmH B 20s 1.3/um M = 5.6
				LmV B 18 0.9/um 5.4
3.	eP	A	12 09 31	<u>Ryukyu Islands</u> 27.47 N 129.29 E
	LmV	B	52.0	H = 11 57 01.7 h = 65 km MB = 4.8
	LmH	B	52.5	D = 84.77 Az = 325.5 (NEIS)
				LmH B 14.5s 0.5/um
				LmV B 16 0.6/um
3.	ePKP	A	21 47 09	<u>Tonga Islands</u> 17.15 S 174.07 W
				H = 21 27 40.8 h = 115 km MB = 4.7
				D = 146.27 Az = 353.5 (NEIS)
				PKPV A 1.8s 60.8nm
4.	eP	A	01 38 04	<u>Kurile Islands</u> 43.19 N 146.95 E
				H = 01 26 03.25 h = normal MB = 5.1
				D = 78.77 Az = 332.9 (NEIS)
				PV A 1.1s 20.2nm M = 5.0
4.	eP	A	12 59 46	<u>Burma</u> 22.64 N 94.06 E
				H = 12 48 50.1 h = 63 km MB = 4.6
				D = 68.13 Az = 317.2 (NEIS)
4.	ePKP	A	15 50 15	<u>New Hebrides Islands</u> 16.50 S 167.10 E
	e	A	50 26	H = 15 30 39.1 h = 9 km MB = 5.3
	LmH	B	16 42.5	D = 140.66 Az = 335.4 (NEIS)
	LmV	B	54.7	LmH B 23s 1.8/um M = 5.7
				LmV B 22 1.6/um 5.7
4.	eP	A	20 11 27	<u>Adriatic Sea</u> 42.16 N 16.17 E
				H = 20 09 12.4 h = 0 km
				D = 9.04 Az = 341 (ISC)

Day	Phase		h m s	Remarks
5.	eP	A	03 55 03.5	Crete 35.44 N 26.45 E
	LmH	B	04 02.1	H = 03 50 51.4 h = 80 km MB = 5.1
	LmV	B	03.8	D = 18.61 Az = 329.3 (NEIS)
				PV A 1.4s 76.6nm M = 4.8
				LmH B 13.5 1.0/um
				LmV B 14 0.8/um
5.	ePg	A	14 59 13	Mining district of the Ruhr
	eiSg	A	59 45.5	Federal Rep. of Germany 51.54 N 7.08 E
				Rockburst
				H = 14 58 12.7 h = 0 km
				D = 3.00 Az = 105.7 (NEIS)
				PgV A 0.8s 9.6nm
5.	eP	A	18 02 48.5	North Atlantic Ridge 52.47 N 31.56 W
	LmH	B	11.9	H = 17 57 11.3 h = normal MB = 4.8
	LmV	B	13.4	D = 26.61 Az = 76.5 (NEIS)
				PV A 1.2s 16.3nm M = 4.5
				LmH B 16 0.8/um 4.4
				LmV B 15 0.9/um 4.6
5.	ePKIKP	A	21 58 35	Solomon Islands 7.45 S 155.93 E
				H = 21 39 33.5 h = 57 km MB = 5.4
				D = 127.74 Az = 332.0 (NEIS)
6.	e(Pn)	A	08 23 36	Yugoslavia 45.3 N 16.0 E
	e	A	24 59	H = 08 21 55 (BCIS)
	eSg	A	25 12.5	D = 6.10
6.	eP1	A	19 56 20.5	Dodecanese Islands 35.24 N 27.73 E
	eP2	A	56 24.5	H = 19 51 55.5 h = 39 km MB = 4.3
				D = 19.33 Az = 327.7 (NEIS)
				P1V A 1.6s 33.0nm M = 4.3
				P2V A 1.6 66.0nm 4.6
7.	ePP	A	05 23 11	Crete 34.3 N 24.8 E
				H = 05 18 41 (BCIS)
				D = 19.55

Day	Phase		h m s	Remarks
7.	e	A	12 30 02	South Atlantic Ridge 23.50 S 13.70 W
	e	A	30 11	H = 12 18 04.1 h = normal MB = 4.4
				D = 77.14 Az = 16.2 (NEIS)
				traces
7.	eP	A	21 19 28	Honshu, Japan 39.33 N 141.95 E
				H = 21 07 21.9 h = 62 km MB = 4.8
				D = 80.37 Az = 330.5 (NEIS)
				PV A 1.0s 11.7nm M = 4.8
7.	e	A	23 45 42	
7.	eP	A	24 07 59	Kodiak Islands Region 56.97 N 152.18 W
				H = 23 56 39.3 h = 48 km MB = 4.3
				D = 71.92 Az = 10.8 (NEIS)
8.	eP	A	01 07 49	El Salvador 13.30 N 89.91 W
				H = 00 55 11.6 h = 88 km MB = 4.9
				D = 86.99 Az = 38.7 (NEIS)
8.	eP	A	02 52 52.5	Unimak Island Region 53.59 N 163.64 W
				H = 02 41 07.9 h = 44 km MB = 4.7
				D = 76.06 Az = 3.1 (NEIS)
8.	eP	A	06 22 52.5	Southern Sumatra 0.25 S 98.40 E
	LmV	B	07 09.0	H = 06 10 03.5 h = normal MB=5.7 MS=5.6
	LmH	B	11.8	D = 88.14 Az = 320.5 (NEIS)
				PV A 1.6s 41.2nm M = 5.5
				LmH B 19.5 1.8 µm 5.5
				LmV B 19 1.9 µm 5.6
8.	eP	A	13 03 40	Talaud Islands 3.63 N 127.12 E
	ePP	A	07 55	H = 12 49 43.6 h = 38 km MB=5.3 MS=4.9
				D = 103.01 Az = 323.9 (NEIS)
8.	LmH	B	14 23.7	Santa Cruz Islands 11.06 S 165.54 E
	LmV	B	34.1	H = 13 09 18 h = 0 km MB = 5.5 (ISC)
				D = 135.2
				LmH B 16s 0.6 µm M = 5.4
				LmV B 18 0.45 µm 5.2

Day	Phase		h m s	Remarks
8.	ePKP	A	14 27 50	New Hebrides Islands 19.02 S 169.40 E
	e	A	28 01.5	H = 14 08 45.0 h = 242 km MB = 5.0
				D = 143.84 Az = 335.9 (NEIS)
				PKPV A 2.0s 42.7nm
8.	eP	A	15 27 30	Afghanistan-USSR Border Region
				36.74 N 71.46 E
				H = 15 19 38.9 h = 170 km MB = 4.0
				D = 44.26 Az = 307.9 (NEIS)
8.	eP	A	19 44 25.5	Turkey 37.36 N 29.49 E
				H = 19 40 05.3 h = 14 km MB = 4.5
				D = 18.42 Az = 321.8 (NEIS)
8.	eP	A	21 07 07	Dodecanese Islands 35.16 N 27.77 E
				H = 21 02 41.4 h = normal MB = 3.7
				D = 19.41 Az = 327.8 (NEIS)
				PV A 1.5s 17.6nm M = 4.1
9.	eP	A	02 45 14	Northwestern Kashmir 35.88 N 73.26 E
				H = 02 36 52.2 h = 33 km MB = 5.0
				D = 45.93 Az = 308.8 (NEIS)
				PV A 1.4s 23.2nm M = 4.9
9.	eP	A	04 47 04.5	Kodiak Island Region 56.91 N 151.81 W
				H = 04 35 43.0 h = 47 km MB = 4.3
				D = 71.94 Az = 11.0 (NEIS)
9.	ePg	A	14 14 54	Switzerland 46.9 N 10.1 E
	eSn	A	15 25	H = 14 13 39 (BCIS)
	eSg	A	15 48	D = 3.9
9.	eP	A	17 53 25	Andreanof Islands, Aleutian Is.
	e(PP)	A	56 45	51.36 N 179.14 W
	e	A	57 10.5	H = 17 41 29.1 h = 48 km MB = 4.8
				D = 77.97 Az = 353.0 (NEIS)

Day	Phase		h m s	Remarks
9.	+eiPKP	AB	20 15 19	New Hebrides Islands 19.91 S 169.75 E
	eX	AB	15 22.5	H = 19 55 45.6 h = 39 km MB=6.2 MS=6.8
	e	AB	15 33	D = 144.79 Az = 335.7 (NEIS)
	e	B	17 19	PKPV A 1.4s 74.4nm
	e	B	20 42	PKPV B 7 7.1/um
	ePS	B	29 03	XV A 1.5 425.0nm
	LmH	B	21 10.0	XV B 9 27.2/um
	LmV	B	13.8	LmH B 26 20.5/um M = 6.7
				LmV B 23 8.6/um 6.5
9.	ePKP	A	21 17 31	New Hebrides Islands 19.77 S 169.88 E
				H = 20 57 57.8 h = 34 km MB=5.9 MS=6.3
				D = 144.71 Az = 335.9 (NEIS)
				PKPV A 1.1s 30.2nm
10.	eP	A	19 40 57	Ecuador 1.44 S 77.74 W
	e	A	41 44	H = 19 28 13.2 h = 181 km MB = 5.3
				D = 90.70 Az = 39.5 (NEIS)
				PV A 1.6s 44.0nm M = 5.2
10.	LmH	B	19 51.9	LmH B 13s 0.8/um
	LmV	B	53.4	LmV B 15 0.45/um
10.	eP	A	21 14 32.5	Southern Iran 27.99 N 56.99 E
				H = 21 06 47.6 h = 14 km MB = 4.6
				D = 40.92 Az = 316.2 (NEIS)
11.	eP	A	00 16 21	Azores Islands 38.74 N 28.67 W
	LmH	B	27.6	H = 00 10 09.2 h = normal MB=5.0 MS=5.0
	LmV	B	27.7	D = 30.64 Az = 53.9 (NEIS)
				PV A 1.8s 40.6nm M = 4.9
				LmH B 14.5 2.3/um 5.0
				LmV B 16 2.9/um 5.1
11.	ePP	A	02 55 09	South of Mariana Islands
	e	A	55 26	13.68 N 146.38 E
	e	A	55 44.5	H = 02 36 38.9 h = 46 km MB=5.5 (NEIS)

Day	Phase		h m s	Remarks
cont. 11.	LmH	B	03 43.8	D = 104.75
	LmV	B	47.3	PPV A 2.0s 51.3nm M = 5.8
				LmH B 17 2.2 µm 5.8
				LmV B 15 1.7 µm 5.7
11.	eP	A	03 53 11	Crete 34.97 N 23.29 E
				H = 03 49 00.1 h = 37 km MB = 4.2
				D = 17.82 Az = 335.1 (NEIS)
				PV A 1.4s 18.6nm M = 4.0
12.	iPn	A	00 03 39	Austria 47.16 N 14.05 E
	ePg	A	03 50.5	H = 00 02 38.1 h = 5 km MB = 4.1
	eSn	A	04 15.5	D = 3.84 Az = 336.2 (NEIS)
	eiSg	A	04 38	PnV A 0.6s 119.0nm
12.	LmV	B	23 10.7	Western Caroline Islands
				6.75 N 137.36 E
				H = 22 02 20.9 h = 28 km MB = 4.9 (ISC)
				D = 106.3
				LmV B 15s 0.55 µm M = 5.3
13.	eP	A	01 59 37.5	Southern Italy 39.4 N 15.5 E
				H = 01 57 02 h = 135 km (BCIS)
				D = 11.65
13.	eP	A	08 11 22	Pyrenees 43.20 N 0.41 W
	eS	A	13 23.5	H = 08 08 39.3 h = 5 km MB = 4.4
				D = 11.09 Az = 43.6 (NEIS)
13.	PKP	A	20 13 04.5	Tonga Islands 15.18 S 173.79 W
	e	A	13 18	H = 19 53 30.1 h = 32 km MB=5.4 MS=5.1
				D = 144.35 Az = 354.1 (NEIS)
				PKPV A 1.4s 30.2nm
14.	+eP	A	03 56 39	Andreanof Islands, Aleutian Is.
				51.32 N 178.30 W
				H = 03 44 43.6 h = 54 km MB = 5.2
				D = 78.06 Az = 353.6 (NEIS)
				PV A 1.7s 39.4nm M = 5.1

Day	Phase		h m s	Remarks
14.	ePKHKP	A	06 19 04	South of Fiji Islands 24.49 S 176.07 W
				H = 05 59 07.9 h = normal MB = 5.1
				D = 153.22 Az = 349.1 (NEIS)
				PKHKPV A 1.3s 17.5nm
14.	iPn	A	07 30 27	Czechoslovakia 49.89 N 13.72 E
	iSg	A	30 48	Explosion
				H = 07 30.0 yield 24 t (PRU)
				D = 1.55 Az = 300 (ISC)
14.	+iP	A	07 54 49.5	Eastern Kazakh SSR 50.04 N 79.01 E
	ePn	A	56 23.5	H = 07 46 57.04 h = 0 km MB=6.0 MS=4.4
	e	A	56 27.5	D = 41.65 Az = 297.8 (NEIS)
				Underground explosion MB = 6.7 (UPP)
				PV A 1.1s 155.0nm M = 5.6
14.	ePn	A	08 00 20	Czechoslovakia 50.76 N 14.42 E
	iSg	A	00 45	Explosion
				H = 07 59.7 yield 25.6 t (PRU)
				D = 1.79 Az = 267 (ISC)
14.	eP	A	09 17 22.5	Caspian Sea 41.88 N 49.03 E
				H = 09 11 46.3 h = 79 km MB = 5.1
				D = 27.07 Az = 301.8 (NEIS)
				PV A 1.0s 21.6nm M = 4.6
14.	+iP	A	17 49 29.5	Andreanof Islands, Aleutian Is.
	LmH	B	18 27.4	51.41 N 177.87 W
	LmV	B	28.5	H = 17 37 35.4 h = 53 km MB = 5.8
				D = 78.01 Az = 353.8 (NEIS)
				PV A 1.5s 191.0nm M = 5.9
				LmH B 18 0.8/um
				LmV B 20 1.3/um
14.	eP	A	18 40 32.5	Ryukyu Islands 28.63 N 129.76 E
	ePP	A	43 57	H = 18 28 20.8 h = 208 km MB = 4.9
				D = 84.06 Az = 325.6 (NEIS)

Day	Phase		h m s	Remarks
15.	eP	A	04 21 58.5	Chagos Archipelago Region 5.36 S 68.60 E H = 04 10 24.2 h = normal MB=5.1 MS=5.0 D = 74.11 Az = 326.3 (NEIS)
15.	eP	A	04 36 18.5	Kodiak Island Region 58.41 N 153.06 W H = 04 25 07.0 h = 52 km MB = 4.8 D = 70.58 Az = 10.3 (NEIS)
15.	eP	A	04 41 38	Chagos Archipelago Region 5.40 S 68.53 E H = 04 30 03.0 h = normal MB = 5.0 D = 74.10 Az = 326.3 (NEIS)
15.	ePKIKP ePKHKP ePKP2	A A A	11 13 31 13 35 13 41	South of Fiji Islands 21.59 S 175.17 E H = 10 54 52.1 h = 567 km MB = 5.3 D = 148.25 Az = 340.0 (NEIS) PKIKPV A 2.4s 82.9nm PKHKPV A 1.6 110.0nm PKP2V A 1.6 82.4nm
15.	eP	A	18 19 03.5	Kurile Islands 45.56 N 151.13 E H = 18 07 01.6 h = 1 km MB = 4.4 D = 77.96 Az = 335.0 (NEIS)
15.	e(PKP)	A	18 49 40.5	New Britain Region 5 S 152 E H = 18 30 29 (NORSAR) D = 123.8
15.	ePKIKP	A	23 16 28	New Hebrides Islands 16.77 S 168.13 E H = 22 57 02.9 h = 7 km MB = 5.6 D = 141.31 Az = 336.1 (NEIS)
15.	eP e LmH	A A C	23 40 51.5 40 56 24 15.8	New Siberian Islands 74.29 N 147.09 E H = 23 31 44.3 h = normal MB = 5.0 D = 51.81 Az = 325.4 (NEIS) PV A 1.7s 60.6nm LmH C 24 0.35 µm M = 4.3

Day	Phase		h m s	Remarks
16.	eP	A	07 33 32.5	Costa Rica 10.94 N 84.49 W H = 07 20 57.1 h = normal MB = 5.1 D = 85.47 Az = 39.4 (NEIS)
16.	eP	A	08 32 16.5	Southern Iran 28.47 N 52.60 E H = 08 25 00.4 h = normal MB = 4.9 D = 37.93 Az = 317.2 (NEIS) PV A 1.0s 9.9nm M = 4.6
16.	eP	A	09 25 09.5	Kashmir - India Border Region 32.28 N 76.03 E H = 09 16 14.1 h = normal MB = 4.9 D = 50.00 Az = 311.4 (NEIS)
16.	eP	A	19 18 20	Southwestern Kashmir 34.18 N 74.05 E H = 19 09 47.7 h = 47 km MB = 5.2 D = 47.51 Az = 310.1 (NEIS) PV A 2.2s 76.4nm M = 5.2
17.	e	A	11 44 05	Northern Italy 44.94 N 7.05 E H = 11 42 04.3 h = normal D = 6.49 Az = 26.7 (NEIS)
17.	+iP LmV LmH	AB B B	22 05 51.0 42.8 44.0	Kurile Islands 48.09 N 154.51 E H = 21 54 02.5 h = normal MB=5.6 MS=5.3 D = 76.60 Az = 336.7 (NEIS) PV A 1.1s 72.5nm M = 5.6 LmH B 19.5 1.2/um 5.1 LmV B 22 1.8/um 5.2
18.	ePKIKP	A	04 27 23	Solomon Islands 5.35 S 154.24 E H = 04 09 08.9 h = 413 km MB = 5.3 D = 125.11 Az = 331.8 (NEIS)
19.	eP	A	03 18 59	Republic of South Africa 27.05 S 26.77 E H = 03 06 58.8 h = normal MB = 5.2 D = 78.51 Az = 350.2 (NEIS)

Day	Phase		h m s	Remarks
19.	eP dif	AC	04 57 20.5	Sumba Island Region 9.44 S 119.45 E
	ePKP	A	01 12	H = 04 43 01.5 h = 58 km MB = 6.0
	ePP	A	01 48	D = 108.54 Az = 320.3 (NEIS)
	LmH	B	05 53.0	LmH B 22s 3.9/um
	LmV	B	58.8	LmV B 20 3.1/um
19.	ePKIKP	AB	13 15 14.5	Fiji Islands Region 20.59 S 176.48 W
	iPKHKP	A	15 19	H = 12 55 57.1 h = 246 km MB = 5.8
	eiPKP2	AZ	15 26.5	D = 149.33 Az = 349.9 (NEIS)
	epPKP	AB	16 16	PKIKPV A 1.5s 80.4nm
	ePP	BC	18 43	PKHKPV A 1.5 679.0nm
	epPP	BC	20 04	
	eSKKS	BC	25 21	
	e	BC	34 00	
19.	ePKP	A	16 52 03.5	Tonga Islands 15.68 S 173.00 W
				H = 16 32 28.5 h = normal MB = 4.9
				D = 144.92 Az = 354.9 (NEIS)
19.	eP	A	18 33 38	Southern Iran 28.15 N 54.98 E
				H = 18 26 09.3 h = normal MB = 4.8
				D = 39.59 Az = 316.7 (NEIS)
19.	e	A	22 38 13	
20.	-iP	AB	17 47 11	Sicily 38.79 N 14.82 E
	ei	A	47 19.5	H = 17 44 25.8 h = 272 km MB = 5.1
				D = 12.07 Az = 350.2 (NEIS)
				PV A 0.8s 292.0nm M = 5.6
21.	eP	A	02 19 14	Tibet 30.29 N 94.87 E
				H = 02 08 47.5 h = normal MB = 4.8
				D = 63.15 Az = 314.9 (NEIS)
21.	-iPn	A	08 19 05.5	Yugoslavia 46.09 N 14.29 E
				H = 08 17 50.9 h = 11 km MB = 4.1
				D = 4.90 Az = 339.7 (NEIS)

Day	Phase		h m s	Remarks
21.	eP	A	09 56 17	<u>Fox Islands, Aleutian Is.</u> 52.30 N 169.38 W H = 09 44 25.3 h = 47 km MB = 4.7 D = 77.43 Az = 359.4 (NEIS)
21.	eP	A	10 03 34.5	<u>Fox Islands, Aleutian Is.</u> 52.24 N 169.36 W H = 09 51 41.3 h = 44 km MB = 4.8 D = 77.49 Az = 359.4 (NEIS)
21.	eP	A	15 40 11	<u>Fox Islands, Aleutian Is.</u> 52.29 N 169.45 W H = 15 28 19.2 h = 49 km MB=4.7 MS=4.6 D = 77.44 Az = 359.3 (NEIS) PV A 1.1s 20.2nm M = 5.1
21.	eP	A	16 10 54	<u>Fox Islands, Aleutian Is.</u> 52.49 N 169.48 W H = 15 59 02.4 h = 49 km MB = 4.7 D = 77.24 Az = 359.3 (NEIS) traces
21.	eP	A	16 12 34	<u>Fox Islands, Aleutian Is.</u> 53.15 N 169.43 W H = 16 00 46.0 h = normal MB = 4.6 D = 76.58 Az = 359.3 (NEIS) traces
21.	ePKHKP	A	17 12 16	<u>Fiji Islands Region</u> 20.54 S 178.77 W
	ePKP2	A	12 22.5	H = 16 53 34.8 h = 606 km MB = 5.0 D = 148.86 Az = 347.2 (NEIS) PKHKPV 1.5s 30.2nm
21.	eP	A	19 24 59.5	<u>Near Coast of Northern California</u>
	e	A	25 07	40.63 N 124.60 W
	LmH	B	20 03.0	H = 19 12 43.5 h = 30 km MB=5.2 MS=4.9
	LmV	B	05.7	D = 81.36 Az = 26.5 (NEIS) LmH B 17s 0.9 µm M = 5.2

Day	Phase		h m s	Remarks
21.	e	A	19 33 42	
21.	LmH	C	23 54.8	<u>West Irian Region</u> 4.30 S 134.02 E H = 22 52 38 h = 19 km (ISC) D = 113.5 LmH C 19s 0.7 /um M = 5.3
22.	epP	A	01 32 37	<u>Near East Coast of Honshu, Japan</u> 35.11 N 140.19 E H = 01 19 59.5 h = 70 km MB = 4.9 D = 83.33 Az = 329.9 (NEIS)
22.	eP	A	02 39 37	<u>Crete</u> 34.12 N 24.82 E H = 02 35 15.0 h = normal MB = 4.0 D = 19.13 Az = 333.7 (NEIS)
22.	e LmH LmV	A B B	14 38 43.5 15 34.3 39.2	<u>Fiji Islands Region</u> 14.81 S 173.18 E H = 14 19 04.1 h = normal MB=5.4 MS=5.3 D = 141.26 Az = 341.2 (NEIS) LmH B 18s 0.8 /um M = 5.5 LmV B 20 0.9 /um 5.5
23.	LmV LmH	B B	20 10.3 10.6	LmH B 20s 0.7 /um LmV B 18 0.5 /um
23.	ePP	A	22 07 32	<u>Crete</u> 35.96 N 25.9 E H = 22 03 09.6 h = 50 km MB = 4.0 D = 17.93 Az = 329 (ISC)
24.	ePKP2	A	07 31 53	<u>Kermadec Islands Region</u> 31.87 S 179.72 W H = 07 11 39.50 h = 200 km MB=4.9 (NEIS) D = 159.44 PKPV A 1.3s 21.8nm
24.	ePKIKP iPKHKP ePKP2 eipPKP	A AB A A	08 33 54 33 58.5 34 03 34 30	<u>Tonga Islands</u> 19.06 S 174.92 W H = 08 14 25.1 h = 121 km MB = 5.7 D = 148.06 Az = 352.1 (NEIS) PKIKPV A 1.6s 49.5nm

Day	Phase		h m s	Remarks
24.	eP	A	13 58 18	Dodecanese Islands 35.05 N 27.82 E
	e	A	58 29	H = 13 53 50.9 h = normal MB = 4.3
	LmV	B	14 06.9	D = 19.53 Az = 327.9 (NEIS)
	LmH	B	07.1	PV A 1.3s 26.2nm M = 4.4
24.	eP	A	18 31 49.5	Near Coast of Peru 12.62 S 77.47 W
	LmV	B	19 14.0	H = 18 18 10.2 h = normal MB = 5.4
	LmH	B	17.6	D = 99.06 Az = 40.1 (NEIS)
				LmH B 18s 0.8 µm M = 5.3
				LmV B 18 0.9 µm 5.3
24.	eP	A	20 27 01	Crete 34.71 N 24.62 E
				H = 20 22 45.4 h = 37 km MB=4.9 MS=5.2
				D = 18.53 Az = 333.2 (NEIS)
				PV A 1.1s 40.3nm M = 4.5
24.	eP	A	20 50 45	North Atlantic Ridge 22.58 N 45.06 W
	e	A	50 51	H = 20 41 37.3 h = normal MB = 4.9
	LmH	B	21 08.6	D = 51.85 Az = 42.6 (NEIS)
	LmV	B	09.0	PV A 1.4s 23.3nm M = 5.0
				LmH B 20 1.1 µm 4.9
				LmV B 20 1.5 µm 5.1
24.	eP	A	21 02 35	Taiwan 24.34 N 121.02 E
	LmH	B	39.7	H = 20 50 08.0 h = 24 km MB = 4.8
	LmV	B	40.7	D = 82.92 Az = 322.8 (NEIS)
				PV A 1.5s 20.1nm M = 5.0
				LmH B 14.5 1.1 µm 5.4
				LmV B 17 0.7 µm 5.2
25.	e	A	00 49 31	Strait of Gibraltar 36.41 N 7.85 W
				H = 00 44 49.9 h = normal MB = 4.5
				D = 19.96 Az = 38.4 (NEIS)
25.	eP	A	02 19 21.5	Kodiak Island Region 56.83 N 153.85 W
				H = 02 07 58.1 h = normal MB = 4.2
				D = 72.22 Az = 9.7 (NEIS)

Day	Phase		h m s	Remarks
26.	eP	A	01 53 16.5	Burma-India Border Region 22.48 N 93.36 E H = 01 42 21.6 h = 50 km MB = 5.1 D = 67.81 Az = 317.1 (NEIS)
26.	eP	A	03 11 28	Mid-Indian Rise 15.23 S 66.90 E
	e	A	11 34	H = 02 59 12.4 h = normal MB = 5.2 D = 81.45 Az = 328.0 (NEIS) PV A 2.4s 55.2nm M = 5.1
26.	ePKIKP	A	18 05 32	South of Fiji Islands 26.85 S 177.73 W
	ePKP2	A	05 57	H = 17 45 59.7 h = 186 km MB = 4.6 D = 155.19 Az = 345.7 (NEIS)
26.	eP	A	20 42 36.5	South of Honshu, Japan 33.40 N 140.83 E
	epP	A	42 53	H = 20 30 06.4 h = 63 km MB = 5.6
	LmH	B	21 20.0	D = 85.08 Az = 330.3 (NEIS)
	LmV	B	28.1	h = 62 km PV A 1.6s 74.1nm M = 5.5 LmH B 17.5 1.0/um LmV B 18 1.1/um
27.	ePP	A	06 06 59.5	South of Mariana Islands
	LmH	B	54.3	13.87 N 146.41 E
	LmV	B	56.4	H = 05 48 42.15 h = 54 km MB=5.1 (NEIS) D = 104.59 PPV A 1.5s 25.1nm M = 5.6 LmH B 16 0.7/um LmV B 16 0.8/um
27.	eP	A	13 38 28.5	Greece 38.64 N 20.36 E
	LmH	B	43.5	H = 13 35 18.6 h = 38 km MB = 4.6 D = 13 51 Az = 335.5 (NEIS) PV A 0.6s 11.5nm M = 4.9 LmH B 16 0.7/um 3.8
28.	ePKP2	A	02 05 54	Tonga 21.3 S 174.0 W H = 01 45 49 h = 0 km MB = 4.6 D = 150.38 Az = 353 (ISC)

Day	Phase		h m s	Remarks
28.	+iPKIKP	AB	05 49 53.4	South of Fiji Islands 23.87 S 179.96 E
	eiPKHKP	A	49 59.5	H = 05 31 06.4 h = 549 km MB = 6.4
	ePKP2	AB	50 11.5	D = 151.78 Az = 344.2 (NEIS)
	epPKP	AB	51 57	PKIKPV A 2.2s 828.0nm
	e(sPKP)	B	53 06	
	e	BC	06 12 22	
	e	B	18 17	
	eSS	BC	22 24	
	esSS	BC	24 52	
28.	ePKIKP1	A	14 01 08.5	New Hebrides Islands 14.46 S 166.60 E
	eiPKIKP2	A	01 14	H = 13 41 45.9 h = 26 km MB=6.4 MS=7.5
	eiPKIKP3	A	01 19	D = 138.62 Az = 336.0 (NEIS)
	ePP	B	03 56	PKIKP2V A 1.4s 97.6nm
	e	B	04 28	PKIKP3V A 1.3 297.0nm
	e	B	04 56	LmH B 20 93.4 /um M = 7.5
	eSP	B	14 32	LmV B 20 12.0 /um 7.6
	eSS	B	22 20	
	LmV	B	15 05.7	
	LmH	B	06.0	
28.	ePKIKP	A	14 38 19	New Hebrides Islands 14.75 S 166.47 E
				H = 14 18 52.0 h = normal MB = 5.6
				D = 138.82 Az = 335.7 (NEIS)
28.	ePKIKP	A	15 06 02.5	New Hebrides Islands 14.66 S 166.87 E
				H = 14 46 33.6 h = normal MB = 5.6
				D = 138.90 Az = 336.1 (NEIS)
28.	ePKIKP	A	15 27 38	New Hebrides Islands 14.49 S 166.47 E
				H = 15 08 12.6 h = normal MB = 5.5
				D = 138.59 Az = 335.9 (NEIS)
28.	ePKIKP	A	16 17 46	New Hebrides Islands 14.61 S 166.99 E
				H = 15 58 18.8 h = normal
				D = 138.90 Az = 336.2 (NEIS)

Day	Phase		h m s	Remarks
28.	ePKIKP	A	16 27 39	New Hebrides 14.67 S 166.74 E
				H = 16 08 21 h = 87 km MB = 5.4
				D = 138.86 Az = 336 (ISC)
28.	ePKIKP	A	18 09 42	New Hebrides Islands 14.92 S 166.77 E
				H = 17 50 15.4 h = normal MB = 5.6
				D = 139.10 Az = 335.9 (NEIS)
28.	ePKIKP	A	18 38 58	New Hebrides Islands 14.95 S 166.50 E
				H = 18 19 30.4 h = 43 km MB = 5.5
				D = 139.02 Az = 335.7 (NEIS)
28.	ePKIKP	A	21 24 59	New Hebrides Islands 14.44 S 166.71 E
	e	A	28 32.5	H = 21 05 36.7 h = 50 km MB = 5.6
	LmV	C	22 34.2	D = 138.64 Az = 336.1 (NEIS)
	LmH	C	34.6	PKIKPV A 1.5s 20.1nm
				LmH C 20 1.4/um M = 5.6
				LmV C 20 1.2/um 5.6
29.	ePKIKP1	AB	00 38 46	New Hebrides Islands 15.12 S 166.90 E
	ePP	BC	42 00	H = 00 19 31.1 h = 47 km MB=6.2 MS=7.2
	ePS	BC	52 12	D = 139.32 Az = 335.9 (NEIS)
	eSS	BC	01 00 20	PKIKPV A 1.3s 78.6nm
	LmH	B	42.7	LmH B 21 64.0/um M = 7.3
	LmV	B	43.3	LmV B 20 60.1/um 7.3
29.	ePKP	A	03 10 08.5	Fiji Islands Region 18.18 S 178.16 W
				H = 02 51 33.7 h = 625 km MB = 4.6
				D = 146.68 Az = 348.7 (NEIS)
29.	ePKIKP	A	07 03 43	New Hebrides Islands 14.85 S 166.51 E
				H = 06 44 18.9 h = normal MB=5.8 MS=5.3
				D = 138.93 Az = 335.7 (NEIS)
				PKIKPV A 1.5s 32.7nm
29.	eP	AB	08 31 45	Komandorsky Islands Region
	e	A	31 57.5	54.64 N 168.74 E
	ePP	B	34 32	H = 08 20 16.2 h = normal MB=5.5 MS=6.0

Day	Phase		h m s	Remarks
cont. 29.	eS	B	08 41 15	D = 73.35 Az = 345.0 (NEIS)
	eSS	B	45 52	PV A 1.5s 47.7nm M = 5.3
	LmH	B	09 13.0	LmH B 17 8.1 μm 6.1
	LmV	B	14.3	LmV B 15 5.9 μm 6.0
29.	ePKIKP	A	09 10 21	New Hebrides Islands 14.25 S 166.54 E
				H = 08 50 49.9 h = 20 km MB = 5.3
				D = 138.40 Az = 336.0 (NEIS)
				PKIKPV A 1.5s 15.1nm
29.	ePKIKP	A	19 33 16	New Hebrides Islands 14.20 S 166.55 E
				H = 19 13 49.8 h = 37 km MB = 5.0
				D = 138.35 Az = 336.1 (NEIS)
				PKIKPV A 1.2s 10.2nm
30.	ePn	A	06 32 30	Central Italy 43.40 N 12.33 E
				H = 06 30 42.0 h = normal MB = 4.1
				D = 7.26 Az = 356.4 (NEIS)
30.	eP	A	10 36 56	Philippine Islands Region
				10.02 N 126.61 E
				H = 10 23 24.2 h = 42 km MB = 5.2
				D = 97.57 Az = 324.4 (NEIS)
30.	ePKP	A	16 41 03	Tonga Islands 15.32 S 173.08 W
				H = 16 21 29.3 h = normal MB=5.4 MS=6.2
				D = 144.56 Az = 354.8 (NEIS)
				PKPV A 1.7s 39.4nm
30.	ePKIKP	A	16 59 02	New Hebrides Islands 15.51 S 166.64 E
	eP'P'	A	17 16 46	H = 16 39 29.7 h = 10 km MB=5.9 MS=6.6
	LmH	C	57.0	D = 139.58 Az = 335.5 (NEIS)
	LmV	C	18 03.0	LmH C 22s 21.9 μm M = 6.7
				LmV C 21.5 39.2 μm 7.1
30.	ePKIKP	A	18 03 16	New Hebrides Islands 15.05 S 166.54 E
				H = 17 43 47.8 h = normal MB = 5.4
				D = 139.13 Az = 335.6 (NEIS)
				PKIKPV A 2.4s 96.6nm

Day	Phase		h m s	Remarks
31.	ePKP2	A	03 03 12	<u>Kermadec Islands</u> 30.01 S 177.46 W H = 02 42 44.12 h = 43 km MB = 4.9 (NEIS) D = 158.29 PKP2V A 1.4s 27.9nm
31.	ePKIKP ePKHKP ePKP2	A A A	03 19 41 19 51.5 20 13	<u>Kermadec Islands Region</u> 28.92 S 178.50 W H = 03 00 12.0 h = 220 km MB = 5.7 D = 156.99 Az = 343.4 (NEIS) PKIKPV A 2.0s 120.0nm PKHKPV A 1.8 81.1nm PKP2V A 1.7 285.0nm
31.	ePKP	A	04 01 13	<u>Samoa Islands Region</u> 15.08 S 172.65 W H = 03 41 40.0 h = normal MB=5.0 MS=4.8 D = 144.35 Az = 355.3 (NEIS) PKPV A 1.3s 28.4nm
31.	eP LmH	A C	11 12 07 46.3	<u>North Eastern China</u> 38.45 N 117.0 E H = 11 01 06 h = 128 km MB = 4.7 D = 69.82 Az = 319 (ISC)
31.	e(PKP)	A	20 16(53)	<u>Fiji Islands Region</u> 17.64 S 178.77 W H = 19 58 12.8 h = 573 km MB = 5.1 D = 146.03 Az = 348.1 (NEIS)

Bei Fragen zur Produktsicherheit wenden Sie sich bitte an:
If you have any questions regarding product safety,
please contact:

Walter de Gruyter GmbH
Genthiner Straße 13
10785 Berlin
productsafety@degruyterbrill.com